Probabilistic Analysis and Related Topics
Volume 2

Contributors

N. U. AHMED
RYSZARD JAJTE
PETER A. LOEB
ARUNAVA MUKHERJEA

Probabilistic Analysis and Related Topics

Edited by A. T. BHARUCHA-REID

DEPARTMENT OF MATHEMATICS
WAYNE STATE UNIVERSITY
DETROIT, MICHIGAN

Volume 2

ACADEMIC PRESS   New York   San Francisco   London   1979
A Subsidiary of Harcourt Brace Jovanovich, Publishers

COPYRIGHT © 1979, BY ACADEMIC PRESS, INC.
ALL RIGHTS RESERVED.
NO PART OF THIS PUBLICATION MAY BE REPRODUCED OR
TRANSMITTED IN ANY FORM OR BY ANY MEANS, ELECTRONIC
OR MECHANICAL, INCLUDING PHOTOCOPY, RECORDING, OR ANY
INFORMATION STORAGE AND RETRIEVAL SYSTEM, WITHOUT
PERMISSION IN WRITING FROM THE PUBLISHER.

ACADEMIC PRESS, INC.
111 Fifth Avenue, New York, New York 10003

United Kingdom Edition published by
ACADEMIC PRESS, INC. (LONDON) LTD.
24/28 Oval Road, London NW1 7DX

Library of Congress Cataloging in Publication Data

Main entry under title:

Probabilistic analysis and related topics.

    Includes index.
    1. Stochastic analysis.   I.   Bharucha–Reid,
Albert T.
QA274.2.P76   519.2   78–106053
ISBN 0–12–095602–0 (v. 2)

PRINTED IN THE UNITED STATES OF AMERICA

79 80 81 82   9 8 7 6 5 4 3 2 1

Contents

List of Contributors   vii
Preface   ix

Optimal Control of Stochastic Systems

N. U. AHMED

I.	Introduction	2
II.	Existence of Optimal Controls without Feedback	3
III.	Existence of Optimal Feedback Controls	12
IV.	Random Necessary Conditions	29
V.	Analytic Necessary Conditions	36
	References	65

Gleason Measures

RYSZARD JAJTE

I.	Introduction	69
II.	Generalities	71
III.	Orthogonally Scattered Gleason Measures	75
IV.	$L_{2\xi}(H)$-Spaces and Isometries Generated by OSG Measures	80
V.	Spectral Gleason Measures	82
VI.	Convergence of Gleason Measures	89
VII.	Gleason Measures in Tensor Products	94
VIII.	Random Gleason Measures	97
	References	103

An Introduction to Nonstandard Analysis and Hyperfinite Probability Theory

PETER A. LOEB

I.	Introduction	105
II.	An Introduction to Nonstandard Analysis	108

III.	A Nonstandard Representation of Measurable Spaces and L^∞	123
IV.	Conversion from Nonstandard to Standard Measure Spaces	128
V.	Applications to Stochastic Processes	135
	References	141

Limit Theorems: Stochastic Matrices, Ergodic Markov Chains, and Measures on Semigroups

ARUNAVA MUKHERJEA

I.	Introduction and Preliminaries	143
II.	Limits of Convolutions in Groups and Semigroups: Analysis in Stochastic Matrices	147
III.	Ergodicity of Markov Chains and Probability Measures on Semigroups: An Interplay	163
IV.	Limit Theorems for Convolution Products of Probability Measures on Completely Simple Semigroups	183
	References	200

Index   205

List of Contributors

Numbers in parentheses indicate the pages on which the authors' contributions begin.

N. U. AHMED (1), Department of Electrical Engineering, University of Ottawa, Ottawa, Ontario, Canada

RYSZARD JAJTE (69), Department of Mathematics, Wayne State University, Detroit, Michigan 48202

PETER A. LOEB (105), Department of Mathematics, University of Illinois, Urbana, Illinois 61801

ARUNAVA MUKHERJEA (143), Department of Mathematics, University of South Florida, Tampa, Florida 33620

Preface

Probabilistic analysis is that branch of the general theory of random functions (or stochastic processes) that is primarily concerned with the analytical properties of random functions. Early research in the field was concerned with the continuity, differentiability, and integrability of random functions. In recent years probabilistic analysis has evolved into a very dynamic area of mathematical research that utilizes and extends concepts and results from functional analysis, operator theory, measure theory, and numerical analysis, as well as other branches of mathematics. The study of random equations is one of the most active areas of probabilistic analysis, and many recent results in the field are due to research on various classes of random equations.

"Probabilistic Analysis and Related Topics," which will be published in several volumes at irregular intervals, is devoted to current research in probabilistic analysis and its applications in the mathematical sciences. We propose to cover a rather wide range of general and special topics. Each volume will contain several articles, and each article will be by an expert in the subject area. Although these articles are reasonably self-contained and fully referenced, it is assumed that the reader is familiar with measure-theoretic probability, the basic classes of stochastic processes, functional analysis, and various classes of operator equations. The individual articles are not intended to be popular expositions of the survey type, but are to be regarded, in a sense, as brief monographs that can serve as introductions to specialized study and research

In view of the above aims, the nature of the subject matter, and the manner in which the text is organized, these volumes will be addressed to a broad audience of mathematicians specializing in probability and stochastic processes, applied mathematical scientists working in those areas in which probabilistic methods are being employed, and other research workers interested in probabilistic analysis and its potential applicability in their respective fields.

Probabilistic Analysis and Related Topics
Volume 2

Optimal Control of Stochastic Systems

N. U. AHMED

DEPARTMENT OF ELECTRICAL ENGINEERING
UNIVERSITY OF OTTAWA
OTTAWA, ONTARIO, CANADA

I.	Introduction	2
II.	Existence of Optimal Controls without Feedback	3
	A. Introduction	3
	B. Linear Systems with Arbitrary Noise	3
	C. Systems Governed by Stochastic Functional Differential Equations	4
	D. Relaxed Stochastic Controls	8
	E. Linear Quadratic Stochastic Control with Random System Coefficients	11
III.	Existence of Optimal Feedback Controls	12
	A. Introduction	12
	B. Continuous Markovian Controls	13
	C. The Separation Principle (Nonanticipative Controls)	14
	D. Partially Observable Linear Diffusion with Discontinuous Controls	17
	E. Functional Differential Systems with Nonanticipative Controls	19
	F. A Class of Degenerate Stochastic Systems	24
	G. Partially Observable Nonlinear Diffusions with Variable Control Constraints	25
	H. Some Open Problems	28
IV.	Random Necessary Conditions	29
	A. Introduction	29
	B. Control of Continuous Diffusions	29
	C. Control of a Class of Jump Parameter Processes	33
V.	Analytic Necessary Conditions	36
	A. Introduction	36
	B. Completely Observable Controller	36
	C. Partially Observable Controller with Past Information	40
	D. Partially Observable Controller with Only Current Information	42
	E. Optimal Policy Involving Controls and Parameters	48
	F. A Class of Degenerate Stochastic Systems	53
	G. A Class of Random Differential Systems with Jump Markov Disturbances	58
	H. Some Open Problems	64
	References	65

I. Introduction

A large number of mathematical models for stochastic systems have been proposed in the literature. A detailed reference to appropriate literature can be found in Bharucha-Reid [17]. In fact every known deterministic mathematical model can be considered as simplification of a suitable stochastic model. Thus there are random algebraic equations, stochastic differential equations, stochastic integral equations, and, more generally, stochastic functional equations. Ordinary differential equations with parameters experiencing Markov jumps, stochastic differential equations combining continuous diffusion and Markovian jump parameters, and stochastic differential equations based on the Wiener process and random Poisson measure have been studied in the literature. Existence and uniqueness of solutions of equations governing these random systems have been studied extensively [3; 17, Chapters 3–7; 15, 32, 34, 35, 65, 74].

The most commonly used model in the study of optimal control theory is the Ito stochastic differential or functional differential equation with Poisson random measure omitted [2, 3, 11, 13, 22, 23, 25, 26, 28, 30–32, 34, 44, 50, 67, 69, 72]. Stochastic differential equations modeled by ordinary differential equations containing Markovian jump parameters have been used in control theory [58, 62, 74]. Recently McShane's model has also been used [7]. Very little is known about the optimal control theory for systems governed by general stochastic functional or integral equations [1, 16, 17]. In contrast, existence theorems for optimal controls of deterministic differential and functional differential equations are well developed [53, 71, 39, 54, 59].

One of the most fascinating and challenging problems in control theory today is the proof of the existence of optimal controls, especially feedback controls, i.e., controls dependent on the state of the system. For systems governed by stochastic differential equations or integral equations of one kind or the other feedback control is more meaningful and also far more complex. In this area the most notable contributions are due to Kushner [44], Wonham [72, 74], Fleming [25, 26, 28, 30, 32], Benés [12, 13], Rishel [56, 58], Varaiya [22, 23, 69], Sworder [62], Duncan [23], Davis [22], Ahmed [1–3, 5, 7], Teo [5, 7], and others.

Other equally important problems are (i) the development of necessary conditions of optimality and (ii) computational methods for determination of the optimal control laws. In Sections II and III we will mainly concern ourselves with the question of existence of optimal controls. In Section II we consider open loop controls and discuss some of the recent existence theorems for systems governed by stochastic differential and functional differential equations, including relaxed stochastic controls. The main tool here is the Prokhorov topology. We have also considered the problem of deparametrization of stochastic control systems. In Section III we consider feedback controls and present some recent existence theorems, including separation principles. Those results admit only continuous controls. In Sections III.C–III.F we discuss existence theorems involving discontinuous controls and functional differential systems. The major tools here are the Girsanov transformation of measures [35], Stroock and Varadhans' martingale theory [65] and generalized implicit function theorems due to McShane, Warfield, Himmelberg, Jacob, and Van Vleck [38, 50].

In Sections IV and V we present recent results on the necessary and sufficient conditions for optimality. Both stochastic necessary conditions, analogous to those for deterministic control problems of Pontryagin type, and analytic necessary conditions involving partial differential equations of either parabolic or hyperbolic type with Cauchy or first boundary conditions are presented.

II. Existence of Optimal Controls without Feedback

A. Introduction

In this section we present existence theorems for optimal random controls in the absence of information feedback, i.e., open loop controls. A great deal of work has been done in this area by several authors like Kushner [44], Fleming and Nisio [32], Baker and Mandrekar [11], Ahmed [2, 3], and others. In the following subsections we will briefly present some of these results.

B. Linear Systems with Arbitrary Noise

Kushner [44] considered the following stochastic integral equation:

$$x(t) = x_0 + \int_0^t A(s)x(s)\,ds + \int_0^t K(u(s))\,ds + z(t) - z(0), \qquad t \in [0, t] \equiv I, \quad (2.1)$$

where x_0 is a fixed n-vector and z is an n-vector random process satisfying

(i) $E\{\max_{t \in I}|z(t)|\} < \infty$, $Ez(t) = 0$, $t \in I$,
(ii) $z(.,.)$ is measurable in the pair (ω, t), where ω is a point in the probability space (Ω, B, P), Ω is the supporting set of the process z, B is the σ-algebra of subsets of the set Ω, and P is the probability measure on B.

For each $t \geqslant 0$ let B_t denote the smallest Borel algebra with respect to which the events $\{z(s): 0 \leqslant s \leqslant t\}$ are measurable. It is clear that B_t is an increasing family of σ-algebras. In case x_0 is assumed to be a random vector the probability space (Ω, B, P) may be considered as the supporting probability space of (x_0, z).

A is an $(n \times n)$ matrix valued Borel measurable function and $K: R^m \to R^n$ is continuous. For admissible controls Kushner introduced the class consisting of all m-vector valued functions $\{u\}$ satisfying

(iii) $u(t, \omega) \in U$ for all $(t, \omega) \in [0, T] \times \Omega$, U compact and convex,
(iv) $u(t, .)$ is B_t-measurable for each $t \in [0, T]$ and $u(., \omega)$ is Lebesgue measurable for each $\omega \in \Omega$.

Denote this class of controls by $\mathcal{U}$. The problem is to find a control $u \in \mathcal{U}$ that minimizes the cost function

$$J(u) = E \int_0^t f_0(x(t), u(t), t) \, dt \qquad (2.2)$$

and satisfies the terminal condition

$$\tilde{g}(Ex(t)) = 0, \qquad (2.3)$$

where $f_0 \, (\geq 0)$ and $\tilde{g}$ satisfy the conditions

(v) $|f_0(x, u, t)| \leq k(1 + |x| + |u|)$,
(vi) $|f_0(x, u, t) - f_0(y, v, t)| \leq k(|x - y| + |u - v|)$, and
(vii) $\tilde{g}$ is uniformly continuous, $k > 0$ fixed.

With this preparation we may now present Kushner's existence theorem [37, Theorem 3, p. 471].

Theorem 2.1 *Consider the system* (2.1) *with A a bounded Borel measurable matrix valued function defined on $I \equiv [0, T]$. Let $K: R^m \to R^n$ be continuous and the set $K(U)$ convex. Assume that the stochastic process z satisfies the conditions* (i), (ii) *and belongs to $C(I, R^n)$ P-a.e., f_0 and $\tilde{g}$ satisfy the conditions* (v), (vi), *and* (vii), *and $J(u) < \infty$ for all $u \in \mathcal{U}$. Then if an admissible control exists satisfying the terminal condition* (2.3), *there is an optimal control $u \in \mathcal{U}$ minimizing* (2.2).

Proof This result is based on Filippov type of arguments as in ordinary control theory. An interesting feature of this result is that the process z is not assumed to be a Wiener process, whereas stochastic differential systems of Ito class are based on Wiener processes.

C. *Systems Governed by Stochastic Functional Differential Equations*

Fleming and Nisio in their early work [25] presented results on the existence of optimal controls for systems governed by stochastic functional differential equations of the form

$$dx(t) = f(t, \pi_t x) \, du(t) + g(t, \pi_t x, \pi_t w) \, dw(t), \qquad t \geq 0,$$

$$x(t) = \hat{x}(t) \quad \text{for } t \leq 0 \quad \text{(a given random initial datum)}$$
$$\pi_t x = \{x(s) : s \leq t\} \quad \text{(the past of } x \text{ up to time } t\text{)},$$
(2.4)

with cost function
$$J(u) = E\phi(x, u), \tag{2.5}$$

where

$f: I \times C(R, R^n) \to R^{n \times m}$ (space of $n \times m$ matrices),

$u: I \to R^m$,

$g: I \times C(R, R^n) \times C(R, R^d) \to R^{n \times d}$,

$\phi: C(R, R^n) \times \mathcal{U} \to R_+$,

and w is a standard d-dimensional Wiener process. The set $\mathcal{U}$ denotes the class of controls, to be defined shortly. The problem is to minimize the cost function $J(u)$ (2.5) subject to the dynamic constraint (2.4). For the sake of comparison we will present some of the major assumptions utilized by Fleming and Nisio.

(F1) f, g are continuous in all their arguments.

(F2) f, g are nonanticipative functionals of Volterra type Lipschitz continuous on $C(R, R^n)$ and $C(R, R^n) \times C(R, R^d)$, respectively, uniformly in $t \in I$.

(F3) the fourth moment of the initial data is bounded.

For details the reader is referred to the original work [32, p. 782]. Let (Ω, B, P) again denote the basic probability space on which both $\hat{x}$ and w are defined. Let D consist of functions $u: I \to R^m$, uniformly Lipschitz continuous on I with a fixed Lipschitz constant and $u(0) = 0$. Let B_t, $t \geq 0$, be an increasing family of σ-algebras with $B_t \subseteq B$.

(F4) the class of admissible controls consists of generalized random variables $\{u\}$ defined on (Ω, B, P) with values in D so that for each $t \in I$, $u(t)$ is B_t-measurable. Denote this class by $\mathcal{U}$.

(F5) The cost functional $\phi \geq 0$ is lower semicontinuous in both its arguments on $C(R, R^n) \times \mathcal{U}$.

Let $\mathscr{F}(C(R, R^n)) \equiv \mathscr{F}((\Omega, B, P), C(R, R^n))$ denote the family of (random variables or measurable functions defined on (Ω, B, P) with values in $C(R, R^n)$ and consider $\chi \subset \mathscr{F}(C(R, R^n))$ to be the family of trajectories $\{x_u : u \in \mathcal{U}\}$, where

$$x_u(t) = \hat{x}(0) + \int_0^t f(\tau, \pi_\tau x_u) \, du(\tau) + \int_0^t g(\tau, \pi_\tau x_u, \pi_\tau w) \, dw(\tau)$$

P-a.e. for each $t \in I$, and $x_u(t) = \hat{x}(t)$, $t \leq 0$, $u \in \mathcal{U}$.

Prokhorov Topology Let Σ_ρ be a separable complete metric space with metric ρ and $\mathcal{B}(\Sigma_\rho)$ the topological Borel field on Σ_ρ. Given two probability measures μ_1 and μ_2 on $\mathcal{B}(\Sigma_\rho)$, the Prokhorov metric on the space of measures $\mathcal{M}(\Sigma_\rho)$ is given by

$$L(\mu_1, \mu_2) = \max\{\delta_{12}, \delta_{21}\},$$

where $\delta_{ij} = \inf\{\delta \geq 0 : \mu_i(A) \leq \mu_j(N_\delta(A)) + \delta\}$, $i, j = 1, 2$, for all closed subsets $A \subset \Sigma_\rho$ and N_δ the δ-neighborhood of A. $(\mathcal{M}(\Sigma_\rho), L)$ is a metric space. Fleming and Nisio established the existence and uniqueness of solutions of the system (2.4) under the assumptions (F1)–(F4) for each control $u \in \mathcal{U}$. Under the given assumptions Fleming and Nisio also proved [32, Theorem 2, p. 787] the following result.

Theorem 2.2 *χ is a sequentially compact subset of $\mathcal{F}(C(R, R^n))$ in the L-topology.*

Fleming then takes an L-closed subset $\mathcal{U}_0 \subset \mathcal{U}$ and proves the following existence theorem.

Theorem 2.3 *If $\phi : C(R, R^n) \times \mathcal{U}_0 \to R$ is lower semicontinuous and $0 \leq \phi(x, u) \leq \infty$, then there exists a control $u_0 \in \mathcal{U}_0$ such that $E\phi(x_0, u_0) \leq E\phi(x, u)$ for all $(x, u) \in \chi \times \mathcal{U}_0$, where x and x_0 are the solutions of the system (2.4) corresponding to the controls u and u_0, respectively.*

The proof of this result is based on lower semicontinuity of the functional $E\phi(x, u)$, which follows from Fatou's lemma, and L-compactness of the set

$$\mathcal{D}(\chi, \mathcal{U}_0) \equiv \{(x_u, u) : u \in \mathcal{U}_0\},$$

which follows from L-compactness of the set χ and the set $\mathcal{U}_0$. The linearity assumption of the local drift f in the control variable u is a limitation on this result. However, this assumption can be dispensed with if f satisfies certain strong continuity conditions with respect to the control variable. Ahmed [3] followed an approach similar to that given by Fleming and Nisio and obtained existence theorems for stochastic functional differential equations described by

$$dx(t) = f(t, \pi_t x, \pi_t u) dt + g(t, \pi_t x, \pi_t w) dw(t),$$
$$t \in I = [0, T], \quad T < \infty, \tag{2.6}$$

where $\pi_t z \equiv \{z(s) : t - a \leq s \leq t\}$, a is a finite positive number, and the initial data is specified over the interval $[-a, 0]$ by giving a second-order random process $\hat{x}$, i.e., $x(t) = \hat{x}(t)$ for $t \in [-a, 0]$. It is assumed that the probability triple (Ω, B, P) now supports all the three given random processes $\hat{x}$, $\hat{u}$, and w. Let B_t, $0 \leq t \leq T$, denote an increasing family of σ-algebras so that

$B_t \subseteq B$ and is independent of the σ-algebras generated by the events $\{w(s) - w(r) : s \geq r \geq t\}$. The problem is to minimize the cost functional

$$J(u) = E\phi(x, u)$$

subject to dynamic constraint (2.6), where

$$\phi : C(I, R^n) \times L_\infty(I, R^m) \to R_+ \equiv [0, \infty). \qquad (2.7)$$

Let U be a nonempty compact convex subset of R^m—and define $D \equiv \{u \in L_\infty(I, R^m) : u(t) \in U \text{ a.e.}\}$. Denote by

$$\mathcal{U} \equiv \mathcal{U}(D) \subset \mathcal{F}((\Omega, B, P), L_\infty(I, R^m))$$

the space of admissible controls with the property that for each $u \in \mathcal{U}$ and for almost all $t \in I$, $u(t,.)$ is B_t-measurable and for P-almost all $\omega \in \Omega$, $u(.,\omega)$ is Lebesgue measurable and belongs to D. The set $D \subset L_\infty(I, R^m)$ with the w^*-topology is compact and closed. This topology is metrizable [24, p. 426] and the metric ρ is the Frechet metric. It has been shown [3, Lemma 1] that the admissible controls $\mathcal{U}(D)$ satisfy the following result.

Lemma 2.1 *The set $\mathcal{U}(D)$ is compact and closed in the L-topology.*

For the proof of existence of optimal controls Ahmed utilized [3, p. 18] the following assumptions:

(A1) *There exists a function $H \in L_2(I \times I', R)$, $I' \equiv [-a, T]$, such that for all $x, y \in C(I', R^n)$ and $u \in D$ and $w \in C(I', R^d)$*

$$|f(t, \pi_t x, \pi_t u) - f(t, \pi_t y, \pi_t u)| \leq \int_{t-a}^{t} |H(t,s)| |x(s) - y(s)| ds,$$

$$|g(t, \pi_t x, \pi_t w) - g(t, \pi_t y, \pi_t w)| \leq \int_{t-a}^{t} |H(t,s)| |x(s) - y(s)| ds.$$

(A2) *There exist $\alpha \in L_2(I, R)$, $\beta \in L_2(I \times I', R)$, and an integer $q \geq 1$ such that for almost all $t \in I$*

$$|f(t, 0, \pi_t u)| \leq |\alpha(t)| \qquad \text{for all } u \in D$$

and

$$|g(t, 0, \pi_t w)| \leq |\alpha(t)| + \int_{t-a}^{t} |\beta(t,s)| |w(s)|^q ds.$$

(A3) $\sup\{E|\hat{x}(t)|^2 : t \in [-a, 0]\} < \infty$.

Using the assumptions (A1)–(A3) Ahmed established [3, Proposition 3, p. 18] the existence and uniqueness of solution of the system (2.6) for each control $u \in D$. It was also shown that the solution x has bounded second moment and that the solution process, as a functional of the control process, is continuous in the L-topology [3, Lemma 2, p. 23] under the assumptions that whenever $u_n \xrightarrow{\rho} u_0, f_x u_n \to f_x u_0$ weakly in $L_2(I, R^n)$, where $(f_x u)(t) \equiv f(t, \pi_t x, \pi_t u)$. Under the above continuity hypothesis Ahmed proved [3, Lemma 3, p. 25] the following closure theorem.

Theorem 2.4 *The set $\chi \equiv \{x_u : u \in \mathcal{U}\}$, where*

$$x_u(t) = \hat{x}(0) + \int_0^t f(\tau, \pi_\tau x_u, \pi_\tau u) \, d\tau + \int_0^t g(\tau, \pi_\tau x, \pi_\tau w) \, dw(\tau)$$

for $t \in [0, T]$ and $x_u(t) = \hat{x}(t)$ for $t \in [-a, 0]$, $u \in \mathcal{U}$, is compact (sequentially) and closed in the L-topology.

With the help of Lemma 2.1 and Theorem 2.4 the following existence theorem was proved.

Theorem 2.5 *If ϕ is a real-valued nonnegative lower semicontinuous functional defined on $\chi \times \mathcal{U}$, then there exists a control $u^* \in \mathcal{U}$ and a corresponding response x^* of the system (2.6) in χ such that $E\phi(x^*, u^*) \leq E\phi(x, u)$ for all $u \in \mathcal{U}$ with $x \in \chi$ being the response corresponding to the control u.*

The crucial assumption used in the construction of the proof of the above result is the continuity of the drift coefficient f in u in the sense defined above. It will be interesting to find broad sufficient conditions that will ensure such continuity. It is quite clear that linearity is a sufficient condition. It should be noted that the form of the dynamical system (2.4) given by Fleming and Nisio appears to be somewhat limited, whereas the form (2.6) is more realistic. The form (2.6) assumes finite memory and (2.4) assumes infinite memory. In (2.4) f and g are assumed to be uniformly Lipschitz in the state variable; this uniformity is not assumed in the form (2.6)

D. Relaxed Stochastic Controls

Recently, Becker and Mandrekar [11] introduced yet another class of stochastic functional differential equations that resembles closely Warga's relaxed dynamical systems [71, pp. 346–352]. This system is described as

$$dx(t) = \int_U f(t, \pi_t x, u) \mu_{t,\omega}(du) + g(t, \pi_t x, \pi_t w) \, dw(t), \qquad t \in I, \qquad (2.8)$$

where μ, for each $t \in I$, is a random measure on the Borel algebra of subsets of the control set U which is assumed compact and convex. The problem is to minimize a performance functional similar to that used in Fleming and Nisio or Ahmed above.

These authors introduced the concept of generalized random controls along the line of Yong [75] and McShane [51] and used Prokhorov's topology to obtain existence theorems for optimal controls. Yong [75] and McShane [51] introduced the class of generalized controls $\mathcal{U}$ as the space of control measures $\{\mu\}$ such that

$$v(h, J) = \int_J \int_U h(t, u) \mu_t(du) \, dt \qquad (2.9)$$

is defined where

(i) for each $t \in J$, μ_t is a measure on the σ-algebra $\mathcal{B}(U)$ of subsets of the set U with $\mu_t(U) = 1$ and for each $A \in \mathcal{B}(U)$, $\mu_t(A)$ is a measurable function on I,
(ii) $h \in C(I \times U, R^n)$,
(iii) $J \subset I$

and $\mu_n \to \mu_0$ finely if for each $h \in C(I \times U, R^n)$, $v_n(h, J) \to v_0(h, J)$ uniformly in J. It is easily shown that this fine convergence induces a metric ρ on $\mathcal{U}$. It is known that $(\mathcal{U}, \rho)$ is a compact separable metric space. The generalized random controls are then defined as follows. Let (Ω, B, P) be the parent probability space and λ a measurable transformation of (Ω, B, P) to $(\mathcal{U}, \rho)$ with

$$\lambda(dt, du, \omega) = \mu_{t,\omega}(du) \, dt,$$

that is,

$$\int_J \int_U h(t, u) \lambda(dt, du, \omega) = \int_J \int_U h(t, u) \mu_{t,\omega}(du) \, dt$$

for all $h \in C(I \times U, R^n)$ P-a.e., where, now, $\mu_{t,\omega}(A)$ is a jointly measurable function of $(t, \omega) \in I \times \Omega$ for all $A \in \mathcal{B}(U)$ and $\mu_{t,\omega}(U) = 1$ for $l \times P$-a.e., where l denotes the Lebesgue measure on the real line and P is the probability measure as defined before. A sequence $\{\lambda_n\}$ of generalized random controls converges to a generalized random control λ if $\rho(\lambda_n, \lambda) \to 0$ P-a.e. This implies that for each $h \in C(I \times U, R^n)$

$$\int_J \int_U h(t, u) \lambda_n(dt, du, \omega) \to \int_J \int_U h(t, u) \lambda(dt, du, \omega)$$

uniformly in J, P-a.e.

Let us denote the class of measure-valued random controls by $(\mathcal{U}, \rho)$. The following result was reported in Becker and Mandrekar [11, Theorem 2.2, p. 4].

Lemma 2.2 *Let I have finite Lebesgue measure. Then the control space $(\mathcal{U}, \rho)$, with elements as measure-valued random functions defined on I, is sequentially compact and closed in the fine topology.*

If the system (2.8) is defined for all $t \in I \equiv [0, \infty)$, then the metric ρ can be redefined as

$$\tilde{\rho}(\lambda_1, \lambda_2) = \sum_{k=1}^{\infty} (1/2^k) \rho_k(\lambda_1, \lambda_2),$$

where ρ_k is ρ defined for measure-valued random functions on I_k each with finite Lebesgue measure and $\bigcup_{k=1}^{\infty} I_k = I$. In this case the control space is denoted by $(\tilde{\mathscr{U}}, \tilde{\rho})$.

Suppose I has finite Lebesgue measure. Let H be a separable Hilbert space and $C(I, H)$ the space of continuous functions on I with values in H. Let $L_2((\Omega, B, P), C(I, H))$ denote the space of random processes $\{x\}$ having the property

$$\int_\Omega \left(\sup_{t \in I} \|x(t)\|_H \right)^2 P(d\omega) < \infty.$$

This defines a metric d on $L_2((\Omega, B, P), C(I, H))$ by

$$d^2(x, y) = \int_\Omega \left(\sup_{t \in I} \|x(t) - y(t)\|_H \right)^2 P(d\omega).$$

The space with this metric is denoted by (C, d). It is easily shown that (C, d) is a complete metric space and that under suitable continuity, Lipschitz, and growth conditions of the drift and diffusion coefficients [11] the stochastic integral operator A defined by

$$(Ax)(t) = \begin{cases} x_0 + \int_0^t \int_U f(s, \pi_s x, u) \lambda(ds, du, \omega) + \int_0^t g(s, \pi_s x, \pi_s w) \, dw(s), & t \in I \\ \hat{x}(t), & t \leq 0 \end{cases}$$

maps $(C, d) \to (C, d)$ and for some integer $q \geq 1$; A^q is a contraction. Using this contraction property, the authors proved the existence and uniqueness of solution of the stochastic integral equation $x = Ax$ (on the Hilbert space H) for each generalized random control $u \in \tilde{\mathscr{U}}$.

It is clear that the solutions are $C(I, H)$-valued random variables. Using Prokhorov topology Becker and Mandrekar [11, Theorem 4.2, p. 20] proved the following result.

Theorem 2.6 *The set of admissible trajectories*

$$\chi \equiv \{x \in (C, d) : x = x_u \text{ for some } u \in \tilde{\mathscr{U}}\}$$

is sequentially compact and closed in the L-topology.

Under the usual lower semicontinuity assumption for the cost functional $J(u) = E\phi(x, u)$, the authors then demonstrate the existence of optimal (generalized random) controls [11, Theorem 5.1, p. 21]. The result of Becker and Mandrekar can be considered as an extension of Warga's existence theorems for relaxed controls of ordinary dynamical systems [70, Theorem 3.1, p. 633; 71, Chapter VI] to stochastic dynamical systems with the state space being a separable Hilbert space. The generalization to Hilbert space only requires careful definition of the corresponding Ito integral.

E. Linear Quadratic Stochastic Control with Random System Coefficients

Recently, Bismut [18] has solved a very general class of linear quadratic finite-dimensional stochastic control problems in which both state and control dependent noise is admitted. The system is governed by the stochastic differential equation

S
$$\begin{cases} dx = (Ax + Cu + f)\,dt + (Bx + Du + g)\,dw \\ x(0) = x_0 \end{cases}$$

and the cost function is given by

$$J(u) = E\left\{\int_0^T |Fx|^2\,dt + \int_0^T (Nu, u)\,dt + |Hx(T)|^2\right\},$$

in which the coefficient matrices A, B, C, D, F, N, H, f, and g are all considered random processes with values in finite-dimensional spaces and w is a finite-dimensional standard Wiener process. Let (Ω, B, P) be a complete probability space supporting $\{w, x_0\}$ with B_t an increasing subsigma field of the σ-field B to which all the random coefficients are adapted. Let β denote the σ-field of well measurable subsets of the set $\Omega \times [0, T]$, V a finite-dimensional innerproduct space, $p, q \in [1, \infty]$, and $L_{p,q}(V)$ the space of $dP \times dt$ classes of β measurable functions with values in V such that

$$E\left\{\left(\int_0^T |f(t)|_V^q\,dt\right)^{p/q}\right\} < \infty.$$

The norm in $L_{p,q}(V)$ is given by

$$\|f\|_{L_{p,q}(V)} = \left(E\left\{\left(\int_0^T |f(t)|_V^q\,dt\right)^{p/q}\right\}\right)^{1/p}$$

where, for $L_{p,\infty}(V)$ the norm is given by

$$\|f\|_{L_{p,\infty}(V)} = \left(P\left(\underset{t\in[0,T]}{\text{ess sup}}|f(t)|\right)^p\right)^{1/p}.$$

Duality brackets are defined between $L_{p,q}(V)$ and $L_{p',q'}(V)$ by

$$E\int_0^T (f(t), g(t))_V\,dt \quad \text{for} \quad f \in L_{p,q}(V) \quad \text{and} \quad g \in L_{p',q'}(V),$$

where $(p)^{-1} + (p')^{-1} = 1$ and $(q)^{-1} + (q')^{-1} = 1$. For any pair of finite-dimensional spaces V and K denote by $M_\infty(V \otimes K)$ the space of essentially bounded nonanticipative β-measurable functions with values in the space of linear operators from V to K. Let $M_\infty^\omega(V \otimes K)$ denote the class of essentially bounded P-measurable functions on Ω with values in $V \otimes K$. Let $C_2^T(V)$

denote the class of right continuous processes $\{x\}$ defined on $[0, T]$ with values in V such that $x(t)$ is B_t-measurable and

$$\|x\| = \left\{ E\left(\sup_{t \in [0,T]} |x(t)|_V^2 \right) \right\}^{1/2} < \infty.$$

Note that $C_2^T(V)$ is a Banach space. Let U be a closed convex subset of a finite-dimensional space and $\tilde{L}_{2,2}(U)$ denote the intersection of the class $L_{2,2}(U)$ and the class of all U-valued processes $\{u\}$ defined on $[0, T]$ such that $u(t)$ is B_t-measurable. Clearly, $\tilde{L}_{2,2}(U)$ is a Hilbert space. The following existence theorem is essentially due to Bismut [18].

Theorem 2.7 *Consider the system* S *with the cost function* J. *Suppose* A, B_i ($i = 1, 2, \ldots m$), $F \in M_\infty(R^n \otimes R^n)$, C, D_i ($i = 1, 2, \ldots m$) $\in M_\infty (U \otimes R^n)$ *with* U *a closed convex subset of* R^r, $N \in M_\infty (U \otimes U)$, $H \in M_\infty^\omega (R^n \otimes R^n)$, $f \in L_{2,1}(R^n)$ *and* $g \in L_{2,2}(R^n \otimes R^n)$. *Then*

(i) *for each control* $u \in \tilde{L}_{2,2}(U)$ *the system has a unique response* $x \in C_2^T(R^n)$ *and*

(ii) *if* $(Nu, u) \geq \alpha |u|_U^2$, *for* $\alpha > 0$, $dP \times dt$-*a.e., then there exists a unique control* $u^* \in \tilde{L}_{2,2}(U)$ *such that* $J(u^*) \leq J(u)$ *for all* $u \in \tilde{L}_{2,2}(U)$.

The first part of the theorem is based on the Banach fixed point theorem and the second part uses the facts that $u \to x$ is an affine continuous mapping from $\tilde{L}_{2,2}(U)$ into $C_2^T(R^n)$, $J(u) \to \infty$ as $\|u\| \to \infty$, and that J is strictly convex and lower semicontinuous on the Hilbert space $\tilde{L}_{2,2}(U)$. For details the reader is referred to Bismut [18, 19], wherein Bismut has also discussed stochastic necessary conditions of optimality. Recently, Bensoussan has considered optimal control problem for the (linear) abstract evolution equation on Banach space [14]. He has also developed a separation principle for the problem [14].

III. Existence of Optimal Feedback Controls

A. Introduction

The results presented in the previous section relate to open loop control problems, that is, where the controllers do not use the information about the past or present states of the process. In many problems of practical interest the controller can utilize the information available about the past states and often produce a more efficient control to be exercised currently. Furthermore, open loop controls require attention from outside the system environment, whereas feedback controls do not.

The study of stochastic feedback controls was initiated by Kushner [44], Wonham [72], and Fleming [26, 28, 30]. Kushner [44] considered optimal control problems of stochastic Ito differential systems with controls that are uniformly Lipschitz in both the state and the time variable, and are allowed to observe the current values of all the components of the state. Wonham [72] considered similar problems with systems linear in the state variable and nonlinear in the controls. The controllers are allowed to use the entire past but only of some of the components of the state variable. Fleming [25] later extended Wonham's result by relaxing the boundedness condition of the control space and weakening certain restrictive conditions on the cost integrand imposed by Wonham. Recently, Fleming considered further generalization of this work and succeeded in developing existence theorems [30] for optimal feedback controls of systems nonlinear in the state variable, but linear in the control with controllers allowed to observe and use partially the current states. Fleming admitted measurable, in contrast to Lipschitzian, controls as considered by Kushner and Wonham. This is a very desirable step forward.

Benés has presented [12] existence theorems of optimal feedback controls for more general systems governed by stochastic functional differential equations of Ito type. In this paper, controllers are allowed complete observation and use of the entire past and are assumed to be merely measurable with respect to an appropriate σ-algebra reflecting the past. However, Benés allows only additive noise, in contrast to Fleming's state dependent noise. This is a limitation of the Girsanov technique and appears to be irremovable except in the special case where the diffusion matrix has an inverse for all space and time variables [22]. Duncan and Varaiya [23] presented an interesting extension of Benés' results by relaxing the growth condition for the drift coefficient. Recently, Ahmed and Teo presented [5] an existence theorem for optimal feedback controls of systems governed by stochastic differential equations of Ito type nonlinear both in the state and the control variables. The controllers are allowed to observe and use partially the current states and are measurable in both the space and time variables. The control constraint set is variable, unlike that in Fleming or Benés. In addition to the existence theorems, Wonham [72, 74] and Fleming [26, 28, 30] among others also developed necessary conditions for optimality from which optimal control laws can be determined. In this section we will briefly present the salient results of Kushner, Wonham, Fleming, Benés, Duncan and Varaiya, Davis and Varaiya, Rishel, and Ahmed and Teo discussed above.

B. *Continuous Markovian Controls*

Early development of stochastic control theory was based on the assumption that the admissible controls were continuous in the state variable. The main reason for imposing this restriction was probably due to lack of results on the existence of solutions of Ito stochastic differential equations with measurable coefficients. In later years, with the availability of results of Stroock and Varadhan [65] and those of Girsanov [35], such restrictions were removed [5, 13, 22, 23, 30]. However, in problems where the cost integrand is known to satisfy a strong convexity condition to be discussed later, the controls turn out to be Lipschitz continuous in the state variable.

C. The Separation Principle (Nonanticipative Controls)

In many problems of practical interest the controller has to make a decision about the controls to be exercised on the basis of information available. This information may be in the form of measurement data modified by the dynamics of the measurement channel and corrupted by noise during transmission and measurement. This problem was first formulated by Wonham [72] for linear systems. He presented a complete solution to this problem, leading to the well-known separation theorem. We present below a brief summary of Wonham's results. For further details the reader is referred to the original papers of Wonham [72, 74].

Let $I = [0, T]$, $T < \infty$, be an interval and $\psi: I \times C(I, R^n) \to R^m$ with the property that if $x_1, x_2 \in C(I, R^n)$ and $x_1(\theta) = x_2(\theta)$ for $0 \leq \theta \leq t < T$, then $\psi(s, x_1) = \psi(s, x_2)$ for all $0 \leq s \leq t$. A functional ψ satisfying the above property is called a nonanticipative Volterra functional. Let Ψ denote the class of nonanticipative Volterra functionals defined on $I \times C(I, R^n)$ with values in $U \subset R^m$ satisfying for all $t, s \in I$ and $y, y_1, y_2 \in C(I, R^n)$ the properties

(i) $|\psi(t, y_1) - \psi(t, y_2)| \leq C_1 \|y_1 - y_2\|$ and
(ii) $|\psi(t, y) - \psi(s, y)| \leq C_2 |t - s|^\alpha$, $\alpha \in (0, 1/2)$, c_i = constant, where U is a compact and convex subset of R^m.

For admissible controls Wonham introduces the class

$$\mathcal{U} \equiv \{u: I \to U : u(t) = \psi(t, \pi_t y) \text{ for some } \psi \in \Psi\},$$

where

$$\pi_t y = \begin{cases} y(s), & 0 \leq s \leq t, \quad y \in C(I, R^n) \\ y(t), & t < s \leq T. \end{cases}$$

The system to be controlled is described by the stochastic differential equation in R^n

$$S_1 \quad \begin{cases} dx(t) = [A(t)x(t) + b(t, u(t))] \, dt + B(t) \, dw_1(t), \quad t \in [0, T] \\ x(0) = x_0 \quad \text{(Gaussian)}. \end{cases}$$

The control $u(t) = \psi(t, \pi_t y)$ for some $\psi \in \Psi$, where y is the response of the information channel governed by the stochastic differential equation in R^n

$$C \quad \begin{cases} dy(t) = F(t)x(t) \, dt + G(t) \, dw_2(t), \quad t \in [0, T] \\ y(0) = 0. \end{cases}$$

It is assumed that all vectors and matrices have real-valued elements with A an $n \times n$ matrix, $b: I \times U \to R^n$, B is an $n \times d_1$ matrix, F is an $n \times n$ matrix, G is an $n \times d_2$ matrix, and w_1 and w_2 are standard Wiener processes with values in R^{d_1} and R^{d_2}, respectively. The problem is to choose a control $u \in \mathcal{U}$, subject to the dynamic constraint S_1, such that the cost functional

$$J(u) \equiv E \int_0^T L(t, x(t), u(t))\, dt \qquad (3.1)$$

is a minimum. Denote by $Y_t \equiv \sigma\{y(s): 0 \leq s \leq t\}$ the σ-algebra generated by the channel output process $\{y(t): t \in I\}$ and define

$$\hat{x}(t) = E\{x(t) \mid Y_t\}$$

where $E\{.\mid.\}$ denotes the conditional expectation and x is the response of the system S_1 corresponding to the control u. Let $\hat{\Psi}$ denote the class of functions $\{\hat{\psi}\}$ defined on $I \times R^n$ with values in U satisfying the property

(i') $\quad |\hat{\psi}(t,\xi) - \hat{\psi}(s,\xi)| + |\hat{\psi}(t,\xi) - \hat{\psi}(t,\eta)| \leq \mathfrak{C}_3(\gamma)|t-s|^\alpha + C_4|\xi - \eta|$

for $0 \leq s \leq t \leq T$, with $\xi, \eta \in S_\gamma \equiv \{x \in R^n : |x| < \gamma\}$, and C_4 independent of γ and $\alpha \in (0, 1/2)$. Let

$$\hat{\mathcal{U}} \equiv \{u: I \to U : u(t) = \hat{\psi}(t, \hat{x}(t)) \text{ for } \hat{\psi} \in \hat{\Psi}\}.$$

It is clear that $\hat{\mathcal{U}} \subset \mathcal{U}$. Wonham presented a complete solution to the problem stated above. He proved the existence of an optimal control and also presented a necessary condition for optimality through the equation of dynamic programming of Bellman that arises from an associated Markovian control problem:

M $\quad \begin{cases} d\hat{x}(t) = [A(t)\hat{x}(t) + b(t, u(t))]\, dt + C(t)\, dw(t), \\ J(u) = E \int_0^T \hat{L}(t, \hat{x}(t), u(t))\, dt = \min, \\ u(t) = \psi(t, \hat{x}(t)) \qquad \text{for } \psi \in \hat{\Psi}, \\ \hat{L}(t, \xi, u) \equiv \int_{R^n} L(t, x, u)\zeta(t, x-\xi)\, dx; \end{cases}$

$\zeta(t, x-\xi)$ is the conditional density of $x(t)$ given $\hat{x}(t) = \xi$ (Gaussian), where C is an $n \times n$ matrix and w is a standard Wiener process in R^n. The complete necessary condition along with an outline of computational procedure will be presented in Section V.C. The significance of Wonham's result lies in the fact that a complex non-Markovian control problem is reduced to a Markovian control problem M and a standard filtering problem. This is known as the "separation theorem." This result is formally presented below after introducing the relevant assumptions on the system and channel parameters. These assumptions are: There exist constants $C_5, C_6, C_7 > 0$, $0 < \beta < \infty$ and $\alpha \in (0, 1/2)$, such that

(W1) The matrices A, B are uhc(α) in t (uniformly Hölder continuous exponent α in t) and F, G are in C' (once continuously differentiable in their domain of definition).

(W2) $G(t)G'(t) \geq C_5 I$ for all t ($I \equiv$ identity matrix).

(W3) $|\det F(t)| \geq C_6$.

(W4) b, b_u, b_{uu} are continuous on $I \times U$ and b, b_u are uhc(α) in t.

(W5) L and L_u are bounded uhc(α) in t and ulc in x (uniformly Lipschitz continuous in x). L_{uu} is bounded and uniformly continuous on $I \times R^n \times U$.

(W6) $(b'(t, u)p + L(t, x, u))_{uu} \geq C_7 I$ for all $(t, x, u, p) \in I \times R^n \times U \times \{p: |p| \leq \beta\}$, β a suitable positive number.

Theorem 3.1 (Wonham [72, Theorem 2.1]) *Under the assumptions (W1)–(W6) an optimal control exists and belongs to the class $\hat{\mathcal{U}} \subset \mathcal{U}$. That is, the optimal control $u^0(t) = \hat{\psi}_0(t, \hat{x}(t))$, $t \in I$, for some $\hat{\psi}_0 \in \hat{\Psi}$.*

The proof of the above result is based on Kalman's filter theory, Bellman's principle of optimality, and an existence theorem for the solution of quasi-linear parabolic partial differential equations (Ladyzhenskaya et al. [49, Theorem 8.1, p. 495]) arising from an application of Bellman's dynamic programming technique to the Markovian problem M. The equation of dynamic programming is given in Section V.C. The most significant fact about Wonham's separation principle is that the optimal control belongs to the class $\hat{\mathcal{U}}$ and that it can be determined by solving the equation of dynamic programming and Kalman's matrix Riccati equation.

It may be noted that the condition (W5) imposed on the cost integrand L is rather strong. Recently, Fleming improved Wonham's separation principle by replacing the conditions (W5) by the conditions (2.2) and (5.1) [25, pp. 212, 223] and removing the boundedness constraint of the control set U and using in its place merely closure and convexity.

Wonham also developed necessary conditions for optimality using the separation theorem. This is discussed in Section V.C. It is essential to mention that the crucial assumptions utilized for developing the separation principle: are: The dynamical system is linear in the state variable, the coefficients are smooth functions of time, and the initial state x_0 is Gaussian.

It is clear that the separation principle greatly simplifies the problem of design of optimal feedback controls. An equivalent separation principle for nonlinear systems does not exist.

D. *Partially Observable Linear Diffusion with Discontinuous Controls*

In many applications cost of storage of past data is prohibitive and therefore controls must be based only on current information available from the measurement data. Such problems may be classified as partially observed Markovian control problems provided also the system is governed by a stochastic differential equation of Ito type. Recently, Fleming presented [30] an existence theorem for optimal controls of partially observed diffusions. The system is described by the Ito stochastic differential equation in R^n

$$S_2 \quad \begin{cases} dx(t) = f(t, x(t), u(t, \tilde{x}(t))) \, dt + g(t, x(t)) \, dw(t), \, t \in [0, T] \equiv I, \\ x(0) = x_0, (\pi_0 - \text{initial probability measure}), \end{cases}$$

where

$$f: I \times R^n \times R^p \to R^n,$$
$$g: I \times R^n \to R^{n \times m} \quad \text{(space of } n \times m \text{ matrices)},$$
$$x = (x_1, x_2, \ldots x_n),$$
$$\tilde{x} = (x_1, x_2, \ldots x_l), \quad 0 \leq l \leq n \quad \text{(observed components)},$$
$$w = (w_1, w_2, \ldots w_m) \quad (m\text{-dimensional standard Wiener process}),$$

and

$$u: I \times R^l \to R^p.$$

For admissible controls Fleming considered the class

$$\mathcal{U}_k = \{u: I \times R^l \to R^p, \text{ measurable, } u(t, \tilde{x}) \in K \text{ a.e.}\},$$

where K is a compact and convex subset of R^p.

Let $B \subset R^n$ be an open set with compact closure, $\bar{B} = B \cup \partial B$, where the boundary ∂B is assumed to be locally representable by functions with Holder continuous second-order partial derivatives to be denoted by $\partial B \in C^{2+\alpha}$, $0 < \alpha < 1$. Define $\Sigma = (I \times \partial B) \cup \{T\} \times B$ to be the terminal set and $\tau = \inf\{t \in I : (t, x(t)) \in \Sigma\}$ the stopping time, that is, the time when the process terminates. Let $L: I \times R^n \times R^p \to R$ be of class C^2. The problem is: Find a control $u \in \mathcal{U}_k$ that minimizes the functional

$$J(u) = E\left\{\int_0^\tau L[t, x(t), u(t, \tilde{x}(t))]\right\} dt, \tag{3.2}$$

subject to the dynamic constraint S_2.

Since $0 \leq l \leq n$, it is clear that the system S_2 also contains the differential equations describing the dynamics of the information channel. In this case the controller is allowed to use only the current information $\tilde{x}(t)$, a part of the current state $x(t)$, and not $\{\tilde{x}(s): s \leq t\}$ as in Wonham. Under usual smoothness conditions of the diffusion coefficient g and the positivity of the matrix $a = \frac{1}{2}gg'$ it is known that for every control $u \in \mathcal{U}_k$ the system S_2 has a unique solution which is a diffusion process. The differential generator of this process is given by

$$A^u\phi \equiv \phi_t + a \cdot \phi_{xx} + f^u \cdot \phi_x$$

where

$$\phi_t = \frac{\partial \phi}{\partial t},$$

$$a \cdot \phi_{xx} = \sum_{i,j=1}^{n} a_{ij}(t, x)\phi_{x_i x_j},$$

$$f^u \cdot \phi_x = \sum_{i=1}^{n} f_i(t, x, u(t, \tilde{x}))\phi_{x_i}.$$

Thus the stochastic optimization problem, considered above, can be shown [67, Theorem 3] to be equivalent to an optimization problem of a system governed by a parabolic differential equation with first boundary conditions:

$$S_2' \qquad \begin{aligned} A^u\phi + L^u &= 0, & (t, x) &\in [0, T] \times B \equiv Q, \\ \phi &= 0, & (t, x) &\in \Sigma, \\ L^u &\equiv L(t, x, u(t, \hat{x})), \end{aligned}$$

with the cost function

$$J(u) = \int_B \phi(0, x) \, d\pi_0(x). \tag{3.2'}$$

The problem is to minimize $J(u)$, defined by (3.2') over $\mathcal{U}_k$, subject to the dynamic constraint S_2'. This problem has been shown to have a solution as stated in the following theorem.

Theorem 3.2 (Fleming [30, Theorem 3, p. 205]) *Let a be uniformly parabolic and belong to C^2, f **linear in u** and measurable in $(t, x) \in Q$, and L continuous in $(t, x) \in Q$ and continuous and convex in the control variable $u \in K$. Then (3.2') has a minimum in $\mathcal{U}_k$.*

The proof of this result is based on the maximum principle for parabolic equations and lower semicontinuity arguments. Under an additional hypothesis it can be shown that the optimal control is continuous. In fact Fleming has shown that if, in addition to the hypothesis as given in the above theorem, L satisfies the so-called strong convexity condition

$$L_{u_i u_j}\beta_i\beta_j \geq b|\beta|^2 \qquad \text{for all} \quad \beta \in R^p, \quad b > 0,$$

then the optimal control satisfies the Holder continuity condition on every compact subset of Q. This result is very useful for practical synthesis of the feedback controller.

Fleming also presented necessary conditions of optimality [30, Theorem 2, p. 202] without the linearity condition of f assumed for the existence theorem. This will be presented in Section V.D. At this point it may be interesting to note the important differences in the results of Wonham and Fleming. In Wonham (i) the system is linear in the state variable x, (ii) the system is not necessarily linear in the control variable u, (iii) the controller is based on the entire past of the observation process, and (iv) x_0 has Gaussian probability density. In Fleming, (i) the system is not necessarily linear in x, (ii) the system is linear in the control u, (iii) the controller is based on the current state of the observation process, and (iv) x_0 has arbitrary probability measure.

E. Functional Differential Systems with Nonanticipative Controls

Before considering optimal control problems, Benés in an earlier paper [13] discussed the problem of existence of optimal strategies based on specified information for a class of stochastic decision problems. The problem may be stated as: Given a continuous stochastic process x on the interval $I = [0, 1]$, find conditions that insure the existence of an admissible strategy $u(t, x)$ minimizing the cost functional $J(u) = E \int_I L(t, x, u(t, x)) \, dt$, where both u and L, at any time t, depend only on the past of the process x up to the present instance t, $0 < t \leq 1$. This result is presented in the following theorem.

Theorem 3.3 (Benés [13, Theorem 1, p. 181]) *Let the process x be continuous on I, P-a.e., where (Ω, B, P) is the probability space on which x is defined, suppose $\int_I E|x(t)|^2 \, dt < \infty$, and let $L(t, x, u)$ be Lebesgue measurable in t, measurable on the Borel sets of $C(I, R^n)$ in x, and continuous in $u \in \Gamma$, Γ a compact metric space with*

$$0 \leq L(t, x, u) \leq k\left(1 + \int_0^t |x(s)|^2 \, ds\right).$$

Then there exists an optimal strategy.

The proof of this result is based on some properties of conditional expectations and a generalized Filippov implicit function lemma due to McShane and Warfield [13, Lemma 1, p. 180; 50, Theorem 1]. Note that here the process x is not affected by the control decision. Recently, Benés [12] presented some interesting theorems for the existence of optimal controls of systems described by Ito stochastic functional differential equations of the type

S₃ $$dx = f(t, x, u(t, x))\, dt + dw, \quad t \in [0, 1] = I,$$

with the cost function

$$J(u) = E \int_0^1 L(t, x, u(t, x))\, dt, \tag{3.3}$$

where f, L, and u are nonanticipative functionals dependent only on the past $\{x(s) : s \leq t\}$ with $f : I \times C(I, R^n) \times \Gamma \to R^n$, $L : I \times C(I, R^n) \times \Gamma \to R$, and $u : I \times C(I, R^n) \to \Gamma$, and Γ is a compact metric space of control values. Benés technique is based on a fundamental result of Girsanov [12, Theorem 1, p. 454] and a version of the implicit function lemma due to McShane and Warfield [12, Lemma 5, p. 460] along with convexity arguments. We will briefly present this result here and refer the reader for details to the original work of Benés [12, 13].

For each $t \in I$, let S_t denote the σ-algebra of events $\{y \in C(I, R^n) : y(s) \in A, 0 \leq s \leq t\}$ with A a Borel subset of R^n. Let $G_t \subseteq S_t$ be a sub-σ-algebra for each $t \in I$. For admissible controls consider the set of functions $u : I \times C \to \Gamma$ which are Lebesgue measurable on I for each $y \in C$ and G_t-measurable on C for each $t \in I$. There is a σ-algebra G over $I \times C$ [12, p. 450] such that admissibility is equivalent to G-measurability. In the work of Benés, Girsanov's result plays a fundamental role and this result is stated below.

Lemma 3.1 (Girsanov [35, Theorem 1, p. 287]) *Let ϕ be any R^n-valued nonanticipative Brownian functional with $\int_I |\phi(t, \omega)|^2\, dt < \infty$, P-a.e., and define*

$$\xi(\phi) = \int_I (\phi, dw) - \tfrac{1}{2} \int_I |\phi|^2\, dt.$$

Then the process $\{\tilde{W}(t) : t \in I\}$, where $\tilde{W}(t) \equiv W(t) - \int_0^T \phi(s, \omega)\, ds$ is a Wiener process under the measure $\tilde{P}$, where $d\tilde{P} = e^{\xi(\phi)}\, dP$ provided $Ee^{\xi(\phi)} = 1$.

Benés has given a martingale proof of this result and has shown that the condition is both necessary and sufficient [12, Theorem 1, p. 454].

The implication of Girsanov's result is that it provides a solution of the stochastic functional differential equation S_3 directly in terms of the measure $\tilde{P}$. Thus the problem of minimization of the cost functional $J(u)$ subject to the dynamic constraint S_3 is reduced to the problem of minimization of the functional

$$J(u) = E\left(e^{\xi(f^u)} \int_0^1 L(t, x, u(t, x))\, dt \right)$$

or

$$J(u) = \int_C \int_0^1 L(t, x, u(t, x)) e^{\xi(f^u)}\, dt\, d\mu(x),$$

where $f^u = f(.,.u(.,.))$ and μ is the measure on $C = C(I, R^n)$ induced by the Wiener process $\{w(t): t \in I\}$. For the existence proof both f and L are assumed to satisfy the following basic conditions:

B(i) $\sigma(t, y, .)$ is continuous on Γ for each $(t, y) \in I \times C$,
B(ii) $\sigma(t, ., u)$ is S_t-measurable for each $(t, u) \in I \times \Gamma$,
B(iii) $\sigma(., y, u)$ is Lebesgue measurable for $(y, u) \in C \times \Gamma$,
B(iv) $|\sigma(t, y, u)|^2 \leq k(1 + \|y(.)\|^2)$, k a constant, for every $(t, y, u) \in I \times C \times \Gamma$,

where σ stands for either f or L. Let D denote the family of attainable densities $D \equiv \{e^{\xi(f^u)}: u \in \mathcal{U}\}$. Then if $G_t = S_t$ i.e., if the whole past is known and if $f(t, y, \Gamma)$ is convex for each $(t, y) \in I \times C$, then D is convex [12, Theorem 3, p. 463]. Benés proved [12, Theorem 4, p. 466] that $L_2 \cap D$ is a strongly closed subset of L_2. Consequently, if D is a bounded subset of L_2, then it is weakly sequentially compact and weakly sequentially closed. From these results Benés proves the following existence theorem.

Theorem 3.4 (Benés [12, Theorem 5, p. 467]) *Let f and L satisfy the assumptions B(i)–B(iii), $G_t = S_t$. Let Γ be a compact metric space, $f(t, y, \Gamma)$ be convex for $(t, y) \in I \times C$, and suppose D is a bounded subset of L_2. Then an optimal control law exists.*

The hypothesis that D is a bounded subset of L_2 is removed by including the condition B(iv). This is given in [12, Theorem 8, p. 471]. In the above results either D (the attainable densities) is assumed to be a bounded subset of L_2 or f is assumed to satisfy also the growth condition B(iv) in addition to the conditions (i)–(iii). The natural place for D is L_1 and so a corresponding weaker version of the existence theorem is possible. This is given in the following result.

Theorem 3.5 (Benés [13, Theorem 6; 7, pp. 470, 471]) *Let f and L satisfy the properties B(i)–B(iv), let Γ be a compact metric space, $G_t = S_t$, and $f(t, y, \Gamma)$ be convex for each $(t, y) \in I \times C$. Then D is both a weakly sequentially closed and weakly sequentially compact subset of L_1 and consequently an optimal control law exists.*

As far as generality is concerned the results given so far in this section can be easily compared. Benés allows the drift coefficient f to be both a nonlinear and nonanticipative functional of both the state and control vectors. In this respect Benés' result is more general than those of Wonham and Fleming. As regards the diffusion coefficient Fleming admits state dependent noise whereas both Wonham and Benés admit merely additive noise perturbing the velocity field. In this respect Fleming's result is more general. Exploiting Girsanov's approach as used by Benés, recently Duncan and Varaiya [23] presented an interesting extension of Benés' results. The significant difference in their results is in the use of a relaxed growth condition for the drift coefficient f as compared to the one used by Benés [B(iv)]. The rest of the conditions are precisely the same. Benés' growth condition is given by

B(iv) $|f(t, x, u)|^2 \leq k(1 + \|x\|^2)$ for all $(t, x, u) \in I \times C \times \Gamma$

and that of Duncan and Varaiya [23] is given by

(DV) $|f(t, x, u)| \leq f_0(\|x\|)$ for all $(t, x, u) \in I \times C \times \Gamma$,

where f_0 is a nonnegative, increasing real-valued function. With this relaxed condition Duncan and Varaiya were able to prove the following fundamental result.

Theorem 3.6 (Duncan and Varaiya [23, Theorem 3, p. 365]) *Suppose f satisfies* B(i), B(ii), B(iii), *and* (DV) *instead of* B(iv), *and let* $f(t, x, \Gamma)$ *be convex, with* Γ *a compact metric space, for each* $(t, x) \in I \times C(I, R^n)$. *Then*

(i) *for an admissible control* $u \in \mathcal{U}$ *there exists a solution to the system equation* S_3 *with continuous sample paths (without explosion) if and only if*

$$Ee^{\xi(f^u)} = 1,$$

(ii) *the set* $D \equiv \{e^{\xi(f^u)} : Ee^{\xi(f^u)} = 1\}$ *of densities is a convex set which is closed in the strong topology of* $L_1(\Omega, B, P)$.

Later on, Duncan and Varaiya presented some sufficient conditions [23, Lemma 9, Corollaries 3 and 4, p. 365] under which the functions $\{f^u\}$ satisfy the condition $Ee^{\xi(f^u)} = 1$. It was shown that Benés' growth condition B(iv) is sufficient. Using Theorem 3.6 the authors proved the following existence theorem.

Theorem 3.7 *Suppose* $\mathcal{U}^0 = \{u \in \mathcal{U} : Ee^{\xi(f^u)} = 1\}$ *is nonempty and let* $L(t, x, u) \equiv L(x)$, $L : C(I, R^n) \to R$, *be bounded and measurable with respect to the σ-algebra of subsets of the set* $C(I, R^n)$. *Then there exists a* $u^* \in \mathcal{U}^0$ *such that*

$$J(u^*) \leq J(u) \quad \text{for all } u \in \mathcal{U}^0,$$

where $J(u) = \int_C L(z) e^{\xi(f^u)} \mu(dz)$ with μ being the Wiener measure on $C = C(I, R^n)$.

The proof of this result is based on the fact that the set $D^0 = \{e^{\xi(f^u)} : u \in \mathcal{U}^0\}$ is a weakly compact subset of $L_1(C, S, \mu)$ and that $J(u)$ is a continuous linear functional of the density $e^{\xi(f^u)}$.

Note that Theorem 3.7 is an extension of Benés' Theorem 3.5 as far as generality of the drift coefficients f^u is concerned. However, in Theorem 3.7 only a subset of the set of admissible controls is allowed. It seems to be an

OPTIMAL CONTROL OF STOCHASTIC SYSTEMS

interesting open problem to find a necessary and sufficient condition for the drift coefficients $\{f^u : u \in \mathcal{U}\}$ that will guarantee the condition

$$E \exp \xi(f^u) = 1 \quad \text{for all } u \in \mathcal{U}.$$

Even though the results of Duncan and Varaiya admit more general drift coefficients, they do not include state dependent noise. Recently, Davis and Varaiya [22, Theorem 2.2, p. 233] have extended this result to include state dependent noise where they consider the system T

$$T \qquad dx = f(t, x, u(t, x)) \, dt + g(t, x) \, dw(t), \quad t \in I = [0, 1],$$

instead of system S_3 with additive noise. In this case additional assumptions are introduced for the diffusion matrix g:

$A_g(\text{i})$ $g(t, z)$ is *nonsingular* for all $(t, z) \in I \times C$, and for fixed $t \in I$, $g(t, .)$ is G_t-measurable, where G_t is an increasing family of sub-σ-algebras of the algebra G of cylinder sets of C as defined earlier.

$A_g(\text{ii})$ $|g^{-1}(t, z) f(t, z, u)| \leq f_0(\|z\|)$ for all $(t, z, u) \in I \times C \times \mathcal{U}$.

Let (Ω, B, P_0) be the probability space on which w is defined and μ_0 be the corresponding Wiener measure on (C, G). Denote this measure space by (C, G, μ_0). Consider the stochastic differential equation

$$dz = g(t, z) \, dw,$$
$$z(0) = z_0 \quad \text{fixed},$$

and let μ denote the measure generated by

$$\mu(F) = \mu_0 z^{-1}(F), \quad F \in G,$$

and (C, G, μ) denote the corresponding measure space.

The set of densities D corresponding to the new system T and the same control class is now given by

$$D = \{\rho(u) : u \in \mathcal{U}\}$$

where

$$\rho(u) = \exp \xi(f^u) \equiv \exp \left\{ \int_I (g^{-1} f^u)' g^{-1} \, dz(t) - \tfrac{1}{2} \int_I |g^{-1} f^u|^2 \, dt \right\}.$$

In this case $\rho(u)$ is regarded as a random variable defined on (C, G, μ) and

$$E\rho(u) = \int_C \rho(u) \mu(dz).$$

Again letting $\mathcal{U}^0 \subset \mathcal{U}$ denote the set of all controls $\{u\}$ for which $E\rho(u) = 1$, one can prove the following result which is essentially due to Duncan and Varaiya and Davis and Varaiya [22, 23].

Theorem 3.8 *Let f satisfy B(i), B(ii), B(iii), and (DV), and let $f(t, x, \Gamma)$ be convex with Γ a compact metric space of control values for each $(t, x) \in I \times C$. Let g satisfy $A_g(\mathrm{i})$ and $A_g(\mathrm{ii})$ and let $L(t, x, u) \equiv L(x)$ be a bounded G-measurable function on C. Then there exists a control $u^* \in \mathcal{U}^0$ such that*

$$J(u^*) \leq J(u) \quad \text{for all} \quad u \in \mathcal{U}^0,$$

where

$$J(u) = \int_C L(z) \exp \xi(f^u) \mu(dz).$$

As before the proof of this result is based on the fact that D is a weakly compact subset of $L_1(C, G, \mu)$ and that J is a linear functional of $\rho(u) = \exp \xi(f^u)$.

It would be of great practical interest to relax the condition $A_g(\mathrm{i})$ and introduce the control in the diffusion matrix g. It appears that even if the condition $A_g(\mathrm{i})$ is retained and control introduced in g, the measure μ in (C, G, μ) is no longer independent of u. Thus $J(u)$ is no longer a linear functional of $\rho(u)$. It is not clear at this time whether the Girsonov approach can be suitably modified to consider the above situation.

F. A Class of Degenerate Stochastic Systems

The Girsanov transformation was also used successfully by Rishel [56, 58] to obtain a necessary and sufficient condition of optimality for a class of systems in which the assumption of positive definiteness of the diffusion coefficient $a = \frac{1}{2}gg'$ is replaced by a weaker assumption which often holds in applications. It is assumed that the system is governed by the following stochastic differential equations:

$$\tilde{R} \quad \begin{cases} dx_1(t) = f^1(t, x(t)) \, dt, \\ dx_2(t) = f^2(t, x(t), u(t)) \, dt + g(t, x_2(t)) \, dw(t), \quad t \in I = [0, T], \\ (x_1(t), x_2(t)) \equiv x(t) \in R^{n_1} \times R^{n_2}, \quad n_1 + n_2 = n, \end{cases}$$

with $gg' > 0$ and

$$J(u) = E \int_0^{\tau_1} L(s, x(s), u(s)) \, ds, \quad \tau_1 = \tau \wedge T,$$

where τ is the first time $(t, x(t)) \in \Sigma = \{T\} \times B \cup [0, T] \times \partial B$. It is also assumed that the solution of the uncontrolled system $(f^2 \equiv 0)$ has a density in $L_p(Q_\delta, R)$, $Q_\delta = [\delta, T] \times B$, $\delta > 0$ for some $p > 1$. Based on these assumptions it is shown that the solution of the system $\tilde{R}$ has a density in $L_{p'}(Q_\delta, R)$ for some $1 < p' < p$ [56, Lemma 1, p. 522]. Using this result (Ito's lemma), Fleming's result on the existence of weak solutions of degenerate quasilinear parabolic partial differential equations [28, 31] and of the McShane–

Warfield implicit function lemma [50, Theorem 1,4], it is shown that along with its partials the weak solution of the equation of dynamic programming corresponding to the system $\tilde{R}$ and the cost function J belongs to $L_q(Q_\delta, R)$, $(q')^{-1} + (p')^{-1} = 1$, and equals the infimum of the cost function J over $\mathcal{U}$. This result also implies the existence of an optimal control provided the existence of a density for the uncontrolled system is assured. For further details the reader is referred to the original literature [56, 58].

G. *Partially Observable Nonlinear Diffusions with Variable Control Constraints*

In an attempt to generalize the results of Fleming on the problem of partially observed diffusions (Fleming [26]) Ahmed and Teo recently showed [5, Theorem 1] that Fleming's assumption of linearity of the drift coefficient f with respect to the control variable can be replaced by some convexity assumptions. Further, it was shown that the control restraint set U can be allowed to be a (set-valued) function of both time and the space variable. The proofs of these results were based on the Filippov technique rather than the lower semicontinuity arguments as used by Fleming. This result will be briefly presented below. For details the reader is referred to the original paper (Ahmed and Teo [5, pp. 351–354]).

The system is again described by S_2 and the cost functional by the expression (3.2). Let $Q \equiv [0, T] \times B$, B an open bounded subset of R^n and $0 \leq T < \infty$. Denote by $\tilde{Q}$ the projection onto $(t, \tilde{x})$-space of the cylinder Q of points (t, x) and let $U(t, \tilde{x})$ be a measurable set-valued function on $\tilde{Q}$ with values that are nonempty compact convex subsets of a fixed compact set $\tilde{U} \subset R^p$. For admissible controls the authors introduced the class $\mathcal{U} \equiv \{u : u \text{ measurable on } \tilde{Q}, u(t, \tilde{x}) \in U(t, \tilde{x}) \text{ for a.a. } (t, \tilde{x}) \in \tilde{Q}\}$. As in Fleming the problem is to minimize the cost function (3.2), subject to the dynamic constraint S_2, by choosing controls from the class $\mathcal{U}$. The proof of the existence theorem to be shortly presented is based on the following assumptions.

(i) $a(t, x) \equiv \frac{1}{2}(gg')(t, x)$ is continuous and bounded on $\bar{Q}$, where $'$ denotes matrix transpose;

(ii) there exists a number $c > 0$ such that $a_{ij}(t, x) z_i z_j \geq c|z|^2$ for all $z \in R^n$ uniformly on $\bar{Q}$ (uniformly parabolic);

(iii) $|a(t, x) - a(\tau, y)| \leq M\{|t - \tau| + |x - y|\}$, where $(t, \tau) \in [0, T]$, $(x, y) \in B$, and M is a positive number; and

(iv) f and L are bounded measurable on $\bar{Q}$ for each $u \in \tilde{U}$ and continuous on $\tilde{U}$ for all $(t, x) \in \bar{Q}$. Further, the set-valued function $\Gamma(t, x)$, $(t, x) \in Q$, defined by

$$\Gamma(t, x) \equiv \{\bar{f}(t, x, u) : u \in U(t, \tilde{x})\}$$

is convex for each $(t, x) \in Q$, where $\bar{f} = \binom{L}{f}$ is the $(n + 1)$-vector constructed by adjoining L to the n-vector f.

Using the assumptions (i)–(iv) for the coefficients a and $\bar{f}$, Ahmed and Teo have shown [5, Lemma 1, p. 353] that for each control $u \in \mathcal{U}$ the system S_2 has a unique solution x_u which is a strong Markov process. The proof of this result is based on some fundamental results of Stroock and Varadhan [65, Theorem 6.2, p. 392, Corollary 3.2, p. 366]. It is shown therein that there exists a unique measure v on S, where S is the σ-algebra of cylinder sets of the space C of continuous functions on I to R^n, so that the process $X_\gamma^z(t), \gamma < t, \gamma, t \in I$, defined by

$$X_\gamma^z(t) \equiv \exp\left\{\langle z, x(t) - x(\gamma)\rangle - \int_\gamma^t \langle z, f\rangle \, d\theta - \tfrac{1}{2}\int_\gamma^t \langle z, az\rangle \, d\theta\right\}$$

is a v-martingale for every $z \in R^n$. The martingale approach solves the Ito equation S_2 and requires that f be only bounded and measurable and a continuous and strictly positive. Further, using Ito's lemma, it has been shown that the optimization problem involving the system S_2 and the cost function (3.2) reduces to the optimization problem of the following first boundary problem [5, Lemma 2; 67, Theorem 1, 3]:

$$S_3' \quad \begin{cases} A^u\phi + L^u = 0, & (t, x) \in Q, \\ \phi = 0, & (t, x) \in \Sigma \end{cases}$$

with the cost function

$$J(u) = \int_B \phi(0, x) \, d\pi_0 = \min, \quad (3.3')$$

where the operator A^u and the function L^u are as defined before for system S_2'.

Lemma 3.2 (Ahmed and Teo [5, Lemma 3, p. 353]) *Let $\bar{f}(t, x, v)$ be a measurable $(n + 1)$-vector-valued function on Q for each $v \in \tilde{U}$ and a continuous function of v on $\tilde{U}$ for each $(t, x) \in Q$. Then, if $\gamma(t, x) \in \Gamma(t, x)$ for all $(t, x) \in Q$ is a measurable function, there exists a measurable function u of $(t, \tilde{x}) \in \tilde{Q}$ with values in $U(t, \tilde{x})$ such that $\gamma(t, x) = \bar{f}(t, x, u(t, \tilde{x}))$ for all $(t, x) \in Q$.*

The proof of this lemma is based on a recent Himmelberg–Jacobs–Van Vleck implicit function theorem [38, Theorem 3', p. 281].

Using the above results, Ahmed and Teo presented the following existence theorem for optimal controls.

Theorem 3.9 (Ahmed and Teo [5, Theorem 1, p. 353]) *Let the system S_3' be given with the cost function J of (3.3'). Suppose a satisfies the assumptions (i)–(iii) and that f and L satisfy the condition (iv). Then there exists a control $u^* \in \mathcal{U}$ that minimizes the cost functional J.*

The proof of this result is based on the argument that the functional J is a continuous linear functional of $\phi \in C(\bar{Q}, R)$ and that the set of trajectories

$$X = \{\phi \in C(\bar{Q}, R) : A^u \phi + L^u = 0, (t, x) \in Q; \phi = 0, (t, x) \in \Sigma, \text{ for some } u \in \mathcal{U}\}$$

is a compact subset of $C(\bar{Q}, R)$. The proof of compactness of the set X is based on (i) the existence of solutions of the first boundary problem S'_3 with measurable coefficients f^u and L^u, (ii) the fact that the solutions of the first boundary problem have the property that $\phi_n \to \phi_0$ uniformly in $C(\bar{Q}, R)$ whenever the coefficients $\bar{f}_n \to \bar{f}_0$ in the w*-topology on $L_\infty(\bar{Q}, R^{n+1})$, (iii) w*-compactness of the set $N = \{y : \bar{Q} \to R^{n+1} : y \text{ measurable and } y(t, x) \in \Gamma(t, x), (t, x) \in \bar{Q}\}$ as a subset of $L_\infty(\bar{Q}, R^{n+1})$, and (iv) convexity arguments using the implicit function Lemma 3.2.

Since the drift coefficient f is permitted to be nonlinear, Theorem 3.9 is an extension of Fleming's result given in Theorem 3.2, where f is assumed to be linear. Yet Theorem 3.9 does not contain Theorem 3.2. However, when time optimal controls are considered or when the cost integrand L is independent of the control variable u Ahmed and Teo's result contains that of Fleming. Further, in Theorem 3.9 the control restraint set is variable, unlike Theorem 3.2. The crucial assumption of Theorem 3.9 is that the set-valued function $\bar{f}(t, x, U(t, \tilde{x}))$ and not merely $f(t, x, U(t, \tilde{x}))$ is convex. This is an undesirable restriction and it may be an interesting problem to remove this constraint. In this respect Cesari–Kuratowski's U-property [21] may be useful. It may be noted that the convexity condition for the set-valued function $\bar{f}(t, x, U(t, \tilde{x}))$ may not be necessary. This is clear from the following result.

Theorem 3.10 *Let the system S'_3 be given with the cost functional J of (3.3') and the control class $\mathcal{U}$ as in Theorem 3.9. Suppose a satisfies assumptions (i)–(iii) and let assumption (iv) be replaced by the assumptions that (L, f) is bounded measurable in $\bar{Q}$ for each $u \in \tilde{U}$ and continuous on $\tilde{U}$ for each $(t, x) \in \bar{Q}$, and that N is a w*-closed subset of $L_\infty(\bar{Q}, R^{n+1})$. Then there exists a control $u^* \in \mathcal{U}$ that minimizes the cost functional J.*

The proof is precisely the same and it uses the fact that a norm bounded w*-closed subset of L_∞ is w*-sequentially compact.

The objective of this section was to present the most up-to-date existence theorems for stochastic control problems admitting measurable coefficients and controls (with the exception of the diffusion coefficient a). If, however, the coefficients a, f^u, L^u are assumed to be sufficiently smooth, say of class C^2, and if f^u and L^u are u.l.c. and the matrices a and $\{L^u_{u_i u_j}\}$ are positive, then optimal controls exist in a smaller class of functions that are u.l.c. also [28, Theorem 1, p. 72].

H. Some Open Problems

Even though a substantial improvement has taken place since 1969 [29] the theory of stochastic control is still far from satisfactory. There are many open questions some of which are presented below.

(i) Extension of the separation principle to a broader class of systems would be of considerable practical interest. One way of treating this problem would be to write the system equations in the desired separated form

$$dx(t) = f(t, x, v(t, \hat{x}(t))) \, dt + g(t, x) \, dw_1(t),$$
$$dy(t) = h(t, x, y) \, dt + k(t, y) \, dw_2(t),$$
$$\hat{x}(t) = E\{x(t) | Y_t\} \equiv (Ty)(t),$$

where

$$(Ty)(t) = \sum_{n=0}^{\infty} \int_0^t \cdots \int_0^t L_n(t, \tau_1, \ldots, \tau_n; dy(\tau_1), \ldots, dy(\tau_n)),$$

$v \in \hat{\Psi} \equiv \{v : I \times R^n \to R^m : v \text{ measurable}, v(t, z) \in U \text{ a.e.}\}$, and prove the existence of a pair (T, v) in some appropriate class that minimizes the cost function

$$J(T, v) = E \int_0^\tau L(t, x, v(t, \hat{x}(t))) \, dt.$$

One main difficulty here is in the representation of the estimator T, which is not a sum of orthogonal operators as in [4, p. 82].

(ii) Admission of measurable diffusion coefficients a. Existing theory of stochastic differential equations admits only local drifts that are measurable functions of both time and space variables [65], whereas the diffusion coefficients are required to be continuous. For bounded measurable diffusions one possible way to tackle this problem may be to construct a sequence of v_n-martingales $\{X_{r,n}^z\}$ corresponding to continuous $\{a_n\}$ (Section II.G) and prove the existence of a measure v such that whenever $a_n \to a$ in the w^*-topology on L_∞, $X_{y,n}^z \to X_y^z$, in some sense, which is a v-martingale. One may require that $\{a_n\}$ be uniformly positive definite.

(iii) Admission of measurable controls in the diffusion coefficients. Kushner's result [44, Theorem 1, p. 465] admits controls in the diffusion coefficient under rather strong assumptions requiring both the coefficients and the controls to be Lipschitz. Solution to problem (ii) will be useful here.

(iv) Extension to general degenerate stochastic systems, not necessarily in the form given by Rishel [56], would be of considerable value in applications.

(v) Admission of control constraints in the form

$$\mathcal{M} = \left\{ u : Q \to R^m : \int_Q |u|^p \, dt \, dx \leq \beta \right\}, \quad p \geq 1, \quad Q \subset I \times R^n, \quad \beta > 0.$$

(vi) Another interesting problem is to find controls that minimize the performance function

$$J(u) = \sup_{A \in B(\Gamma)} |Q^u(T,A) - Q^*(A)|,$$

where $Q^u(T,.)$ is the measure induced on R^n by the solution process at a given instant T corresponding to the control u, and Q^* is a desired probability measure with Γ being the supporting set and $B(\Gamma)$ the Borel subsets of the set Γ.

(vii) For other interesting open problems see questions 13.4, 13.5 in Fleming [29, pp. 504–505].

IV. Random Necessary Conditions

A. Introduction

In the rest of this article we present a brief survey of recent developments in the area of necessary conditions for optimality for stochastic dynamical systems. Two distinct lines of development have taken place. One is concerned with pointwise necessary conditions similar to those of Pontryagin, which give random adjoint differential equations and conditional expectation of random Hamiltonians from which random controls are to be determined. We discuss this approach in the present section under the name random necessary conditions. The second line of development is based on Ito's lemma and Bellman's principle of optimality, which give a quasilinear parabolic differential equation with Cauchy condition for fixed time problems or with first boundary condition for Markov time problems. The solution of the associated Cauchy or first boundary problem provides sufficient information to determine the optimal feedback control law. We discuss this approach in Section V under the name analytic necessary conditions.

From a computational point of view the analytic necessary conditions appear to be more promising than the random necessary conditions.

B. Control of Continuous Diffusions

Stochastic versions of Pontryagin's maximum principle have been extensively investigated in the literature; see, e.g., Florentine [33], Kushner [42–48], and Sworder [61–64]. These necessary conditions resemble strongly those of Pontryagin.

In a series of papers [42–45, 47] Kushner has developed necessary conditions for optimality in which the adjoint system giving the stochastic Lagrange multiplier is precisely the same as that in the deterministic problem.

The Hamiltonian is replaced by its conditional expectation conditioned upon the σ-algebra generated by the observable process. The system is governed by a stochastic differential equation of the form

$$dx(t) = f(t, x(t), u(t)) \, dt + dz(t), \quad t \in I,$$

where z is not necessarily a Wiener process and u is a control process satisfying certain admissibility conditions. For this problem the reader is referred to [41]. We consider here a more recent result of Kushner [42] that involves state constraints. It is a well-known fact that necessary conditions for optimality with state constraints are rather difficult even for deterministic systems [54, Theorem 22, p. 267]. In a recent paper Kushner has developed some very interesting necessary conditions for optimality with terminal and intermediate state constraints for stochastic systems governed by the Ito differential equation. These results are proved using the powerful abstract variational theory developed by Neustadt [52]. It will be difficult, if not impossible, to incorporate such state constraints using the method of dynamic programming. We will briefly present here this result and refer the reader for details to Kushner's original work [42].

Consider the system governed by the Ito stochastic differential equation in R^{n+1}

$$dx(t) = f(t, x(t), u(t)) \, dt + g(t, x(t)) \, dw(t), \quad t \in [0, T] \equiv I, \quad (4.1)$$
$$x(0) = \text{variable} \quad \text{(determined by terminal constraint)},$$

where

$x(t) = (x_0(t), x_1(t), \ldots, x_n(t))'$ ($'$ denotes transpose),
$f: I \times R^{n+1} \times R^m \to R^{n+1}$
$g: I \times R^{n+1} \to R^{(n+1) \times d}$ ($(n+1) \times d$ matrices),
$w = (w_1, w_2, \ldots w_d)$ (a d-dimensional standard Wiener process).

Note that integral costs are absorbed in the dynamical system (4.1) in its first component. Let $0 = t_0 < t_1 < t_2 \cdots < t_{k+1} = T$ be a set of fixed points in the interval I, let $n_0, n_{k+1}, m_0, m_1, \ldots, m_{k+1}$ be given positive integers, and let

$$\tilde{\gamma}_j : R^{n+1} \times R^{n+1} \to R^{n_j}, \quad j = 0, k+1$$
$$\tilde{q}_j : R^{n+1} \times R^{n+1} \to R^{m_j}, \quad j = 0, 1, 2, \ldots k+1,$$

and

$$h: R^{n+1} \to R$$

be a given set of Borel functions defined on their respective domains. Let $U(t)$, $t \in I$, be a set-valued map on I whose values are nonempty closed convex subsets of the space R^m. Let (Ω, B, P) be the probability space on which w is defined, $\{B_t : t \in I\}$ an increasing family of subsigma algebras contained in B which are nonanticipative with respect to the Wiener process w. Let $\mathcal{U}$ denote the class of admissible controls $\{u\}$ on I with values $u(t) \in U(t)$ for $t \in I$ such that $u(t)$ is B_t-measurable.

The problem is to find a control $u^* \in \mathcal{U}$ such that

(i) it minimizes the cost function

$$J(u) = E\{x_0(T) + h(x(T))\}, \qquad (4.2)$$

(ii) the corresponding trajectory denoted by x^* satisfies the Ito equation

$$\{dx(t) = f(t, x(t), u^*(t))\, dt + g(t, x(t))\, dw(t), \qquad t \in I, \qquad (4.3)$$

and

(iii) the trajectory x^* satisfies the terminal and intermediate constraints

$$\begin{aligned}\gamma_0(x^*(0)) &\equiv E\{\tilde{\gamma}_0(x^*(0), Ex^*(0))\} = 0, \\ \gamma_{k+1}(x^*(T)) &\equiv E\{\tilde{\gamma}_{k+1}(x^*(T), Ex^*(T))\} = 0,\end{aligned} \qquad (4.4)$$

and

$$q_i(x^*(t_i)) \equiv E\{\tilde{q}_i(x^*(t_i), Ex^*(t_i))\} \leqslant 0 \qquad (4.5)$$

for $i = 0, 1, 2, \ldots k + 1$. (The convention $z \leqslant 0$ is used to denote that each component of the vector z is nonpositive).

Let us write $\gamma_{i,x}^*$, $\gamma_{i,e}^*$ [matrices of dimension $n_i \times (n+1)$], $i = 0, k+1$, and $q_{i,x}^*$, $q_{i,e}^*$ [matrices of dimension $m_i \times (n+1)$], $i = 0, 1, \ldots k+1$, for the matrices of first partial derivatives of $\tilde{\gamma}_i(\alpha, \beta)]$ $(i = 0, k+1)$ and $\tilde{q}_i(\alpha, \beta)$ $(i = 0, 1, \ldots k+1)$ with respect to the first and second arguments (α and β) evaluated at $\alpha = x^*(t_i)$ and $\beta = Ex^*(t_i)$, respectively. Let h_x^* denote the gradient of h evaluated at $x^*(T)$.

Let x^* and u^* denote the optimal pairs (which are assumed to exist) and consider the following Ito stochastic linear differential equations:

$$dy(t) = f_x^* \cdot y(t)\, dt + \left(\sum_i dw_i\, g_{i,x}^*\right) y(t), \qquad (4.6)$$

and

$$d\Phi(t, \tau) = f_x^* \Phi(t, \tau)\, dt + \left(\sum_i dw_i\, g_{i,x}^*\right) \Phi(t, \tau) \qquad (4.7)$$

for $0 \leqslant \tau \leqslant t \leqslant T$, where $f_x^* \equiv \operatorname{grad}_x f(t, x^*(t), u^*(t))$, $g_{i,x}^* \equiv \operatorname{grad}_x g_i(t, x^*(t))$, and g_i is the ith column of the diffusion matrix g. Both the equations (4.6) and (4.7) have unique continuous (in t) solutions with probability one with finite second moments. Assuming that

$$\Phi(\tau, \tau) = I, \tag{4.8}$$

the solution of (4.6) can be expressed as

$$y(t) = \Phi(t, \tau) y(\tau), \quad 0 \leq \tau \leq t \leq T, \tag{4.9}$$

starting from the initial state $y(\tau)$, where the equality holds with probability one. It is known [42] that the random transition matrix Φ corresponding to the random linear system (4.6) satisfies the usual semigroup property

$$\Phi(t, s)\Phi(s, \tau) = \Phi(t, \tau) \quad \text{with probability one,} \tag{4.10}$$

where $t \geq s \geq \tau \geq 0$ and Φ has a continuous (in t) representative. For the proof of the necessary condition of optimality this fact has been used in addition to a number of technical assumptions [42, Theorems 2.1–2.5, p. 552] that include:

(i) $x(0)$ is B_0-measurable and has finite second moment;
(ii) f, g are Borel functions having uniformly bounded first partials and satisfying the growth condition

$$\{|f(t, x, u)|^2, |g(t, x)|^2\} \leq k_0(1 + |x|^2)$$

independently of t and u, for a suitable constant $k_0 > 0$;
(iii) $\tilde{\gamma}_i$ ($i = 0, k + 1$), $\tilde{q}_i$ ($i = 0, 1, 2, \ldots, k + 1$), and h are Borel functions having square integrable first partials and satisfying the growth conditions

$$\{|\tilde{\gamma}_i(\alpha, \beta)|, |\tilde{q}_i(\alpha, \beta)|\} \leq k_0\{1 + |\alpha|^2 + |\beta|^2\} \quad \text{and} \quad |h(\alpha)| \leq k_0\{1 + |\alpha|^2\}.$$

The necessary conditions for optimality are given in the following theorem.

Theorem 4.1 *Suppose the given assumptions are satisfied. Then in order that u^*, x^* be an optimal pair it is necessary that there exist a nonzero $(n + 1)$-dimensional stochastic process $\{\psi(t): t \in I\}$ (or equivalently a scalar $\theta \leq 0$ vectors $\alpha_i \leq 0$, $i = 0, 1, 2, \ldots, k + 1$, $\alpha_i = 0$ if $q_i(x^*(t_i)) < 0$, and vectors β_i, $i = 0, k + 1$) and a set $\tilde{I} \subset I$ of Lebesgue measure zero so that for all $t \in I \setminus \tilde{I}$ and B_t-measurable $U(t)$-valued random variables u_t and admissible initial state $x(0)$ the following inequalities hold with probability one:*

(i) $E\psi'(t)[f(t, x^*(t), u_t) - f(t, x^*(t), u^*(t))] \leq 0,$
(ii) $E[\psi'(0) + \alpha_0'(q_{0,x}^* + Eq_{0,e}^*) + \beta_0'(\gamma_{0,x}^* + E\gamma_{0,e}^*)] \delta x(0) \leq 0.$

Further, with probability one,

(i') $E\{\psi'(t)[f(t, x^*(t), u_t) - f(t, x^*(t), u^*(t))] | B_t\} \leq 0,$
(ii') $E\{[\psi'(0) + \alpha_0'(q_{0,x}^* + Eq_{0,e}^*) + \beta_0'(\gamma_{0,x}^* + E\gamma_{0,e}^*)] | B_0\} = 0,$

where

$$\psi'(t) = \begin{cases} \psi'(T)\Phi(T,t) & \text{for } t \in [t_k, T] \\ [\psi'(t_i) + \alpha'_i(q^*_{i,x} + Eq^*_{i,e})]\Phi(t_i, t) & \text{for } t \in [t_{i-1}, t_i) \end{cases}$$

for $i = 1, 2, \ldots k$, and

(iii') $\psi'(T) \equiv \{\theta(e_1 + h^*_x)' + \beta'_{k+1}(\gamma^*_{k+1,x} + E\gamma^*_{k+1,e}) + \alpha'_{k+1}(q^*_{k+1,x} + Eq^*_{k+1,e})\}$ with $e_1 = (1, 0, 0, \ldots 0)'$ an $(n+1)$-vector.

If the diffusion matrix $g \equiv 0$, the above theorem reduces to the Pontyragin maximum principle. For example, if, with probability one, the system starts from a fixed state $x_0 \in R^{n+1}$ and if all the state constraints are removed then the necessary conditions given above reduce to

$$E\psi'(t)[f(t, x^*(t), u_t) - f(t, x^*(t), u^*(t))] \leq 0,$$
$$d\psi(t) = -f'_x(t, x^*(t), u^*(t))\psi(t)\,dt - \sum_i dw_i\, g'_{i,x}(t, x^*(t))\psi(t) \qquad (4.11)$$
$$\text{with} \quad \psi(T) = \theta[e_1 + h_x(x^*(T))],$$
$$dx^*(t) = f(t, x^*(t), u^*(t))\,dt + \sum_i dw_i g_i(t, x^*(t))$$

with $x^*(0) = x_0$

and, further, if $g \equiv 0$, one has the Pontryagin maximum principle for fixed terminal time and free end conditions.

Remark The usefulness of the maximum principle (Theorem 4.1) for computation of optimal controls is not known. Suppose all the intermediate constraints are removed, leaving only the terminal constraints (ii') and (iii). Following the usual procedure from deterministic control theory, the inequality (i) or (i') may be used to obtain the form of the optimal control. If this control is then substituted in the system equation (4.1) and the adjoint equation corresponding to (4.6), one obtains a stochastic two-point boundary value problem involving an Ito stochastic differential equation in the variables (x, ψ) in $R^{2(n+1)}$ with terminal constraints (ii') and (iii). The author is not aware of any direct method for solving stochastic two-point boundary value problems.

C. *Control of a Class of Jump Parameter Processes*

Recently, Sworder [62, 64] has developed necessary conditions of optimality for a class of linear systems in which both the plant and control matrices are jump Markov processes with finite state space. The system is described by the random differential equation in R^n

$$\frac{dx}{dt} = A(t)x(t) + B(t)u(t), \qquad t \in I = [0, T],$$

$$x(0) = x_0$$

for which $[A, B]$ is a separable Markov process on I with values in a finite subset $K \subset M \times N$, where M and N denote the class of $(n \times n)$ and $(n \times m)$ matrices, respectively, with entries $\{a_{ij}, b_{ij}\}$ real. Suppose K consists of s elements from $M \times N$. Let $Q(t) = \{q_{ij}(t)\}$, $t \in I$, $i, j = 1, 2, \ldots s$, denote the matrices of transition probabilities of the plant and control parameters $[A, B]$, so that for $i, j = 1, 2, \ldots s$

$$\begin{aligned}\text{Prob}\{[A(t + \Delta t), B(t + \Delta t)] &= [A_j B_j] | [A(t), B(t)] = [A_i, B_i]\} \\ &= \begin{cases} q_{ij}(t)\Delta t + o(\Delta t) & \text{for } j \neq i \\ (1 + q_{ij}(t)\Delta t) + o(\Delta t) & \text{for } j = i. \end{cases}\end{aligned} \qquad (4.12)$$

Let $p = (p_1 p_2 \cdots p_s)'$ denote the initial probability vector, so that

$$\text{Prob}\{[A(0), B(0)] = [A_i B_i]\} = p_i, \qquad i = 1, 2, \ldots s.$$

Let $\{\gamma(t) : t \in I\}$ be a stochastic process in R^k representing measurements of the observable and physically "measurable" components of the process $\{[A(t), B(t)], t \in I\}$. Let V denote the class of admissible controls consisting of continuous bounded functions defined on $I \times R^n \times R^k$ with values $v(t, x(t), \gamma(t)) \in R^m$ and C^1 in its second argument.

Under the stated assumptions it can be verified that for each $v \in V$ and $x_0 \in R^n$ the equation

$$\frac{dx}{dt} = A(t)x(t) + B(t)u(t), \qquad t \in [0, T],$$

$$x(0) = x_0, \qquad (4.13)$$

$$u(t) = v(t, x(t), \gamma(t))$$

has a unique solution absolutely continuous on I with probability one. The problem is to find a control law $u^* \in V$ that minimizes the cost function

$$J(v) = E\left\{\int_0^T L(t, x(t), v(t, x(t), \gamma(t))) \, dt\right\}. \qquad (4.14)$$

For this, define for each control law $u \in V$ the function $H: I \times R^n \times R^n \times R^m \to R$ by

$$H(t, x, \psi, u) = \psi'(A(t)x + B(t)u(t, x, \gamma(t))) - L(t, x, u(t, x, \gamma(t))). \qquad (4.15)$$

Let v^* be the optimal control law and x^* the corresponding response of the system (4.13) with $u(t) = v^*(t, x^*(t), \gamma(t))$. Let ψ^* be a solution of the random differential equation

$$\frac{d\psi}{dt} = -(A(t) + B(t)v_x^*(t, x^*(t), \gamma(t)))'\psi + [L_x(t, x^*(t), v^*(t, x^*(t), \gamma(t)))$$
$$+ L_v(t, x^*(t), v^*(t, x^*(t), \gamma(t)))v_x^*(t, x^*(t), \gamma(t))]' \quad \text{with} \quad \psi(T) = 0.$$
(4.16)

The necessary condition for optimality can then be stated as follows:

Theorem 4.2 *For (v^*, x^*) to be an optimal pair it is necessary that the inequality*

$$E\{H(t, y, \psi^*(t), v^*(t, y, z)) | x^*(t) = y, \gamma(t) = z\}$$
$$\geq E\{H(t, y, \psi^*(t), v(t, y, z)) | x^*(t) = y, \gamma(t) = z\}$$

be satisfied with probability one for almost all $t \in I$ and for all $v \in V$.

In other words, the optimal control law maximizes the conditional expectation of the stochastic Hamiltonian. It turns out that if $\{\gamma(t): t \in I\}$ is a Markov process and if $[A(t), B(t)]$, $t \geq \tau$, is conditionally independent of $[x(l), \gamma(l)]$, $l < \tau$, given $[x(\tau), \gamma(\tau)]$, then the above necessary condition is also a sufficient condition. Using this result the linear regulator problem with quadratic cost

$$L(t, x, v) = x'Rx + v'Sv \quad (4.17)$$

can be satisfactorily resolved provided that the observed process $\{\gamma(t): t \in I\}$ consists of all the components of the system parameters $\{[A(t), B(t)]: t \in I\}$. Suppose both R and S are symmetric and real and $R \geq 0$ and $S > 0$. Using the necessary conditions, it is then easily verified that the optimal control law is of the form

$$u = \tfrac{1}{2}E\{S^{-1}B'(t)\psi(t) | x^*(t) = x, [A(t), B(t)] = [A_i \; B_i]\}, \quad (4.18)$$

where ψ is the solution of the differential equation

$$d\psi(\tau)/d\tau = -(A(\tau) + B(\tau)v_x')\psi(\tau)$$
$$+ 2(Rx + v_x'Sv), \tau \in [t, T] \quad \text{with} \quad \psi(T) = 0. \quad (4.19)$$

Clearly, $\psi(\tau), \tau \in [t, T]$, is conditionally independent of $\{x^*(\tau): \tau < t\}$ given $x^*(t) = x$. Thus if we choose $\psi(t)$ to be linear in x, (4.18) and (4.19) remain compatible. Let $\{K(t): t \in I\}$ be an $(n \times n)$ matrix-valued stochastic process such that for $t \in I$, $K(t)$ is conditionally independent of $\{\gamma(\tau): \tau > t\}$ given $\gamma(t)$. Then choosing

$$\psi(t) = 2K(t)x \quad (4.20)$$

and defining
$$K_i(t) = E\{K(t)|[A(t),B(t)] = [A_iB_i]\}, \qquad (4.21)$$
one obtains
$$v(t,x,i) = S^{-1}B_i'K_i(t)x, \qquad i = 1, 2, \ldots, s. \qquad (4.22)$$

Using the defining relations (4.12), (4.20), (4.21), and the expression (4.22) in the differential equation (4.19), one obtains a differential equation for K_i:

$$dK_i/dt = -K_i(t)A_i - A_i'K_i(t) - K_i(t)[B_iS^{-1}B_i']K_i(t)$$
$$+ R - \sum_{j=1}^{s} q_{ij}(t)K_j(t), \qquad t \in [0,T),$$
$$K_i(T) = 0, \qquad i = 1, 2, \ldots, s. \qquad (4.23)$$

Existence and uniqueness of solutions of the system (4.23) is well known [20, p. 66; 74, p. 193]. Thus the equations (4.22) and (4.23) provide a complete synthesis of a unique feedback control law.

V. Analytic Necessary Conditions

A. Introduction

In this section we intend to present in some detail the major developments in the area of necessary conditions for optimality that employ analytic techniques involving partial differential equations rather than the probabilistic approach relying on the stochastic maximum principle. The significant advantage of the analytic approach lies in the computability of the optimal control law from the solution of some quasi-linear parabolic partial differential equations or, in some cases, quasi-linear parabolic integropartial differential equations. Several interesting results in this area are due to Fleming [25–30], Wonham [72–74], Rishel [56, 57], and Ahmed and Teo [5–7]. We wish to present a systematic account of these results. Furthermore, each result will be followed by a list of computational steps that may be used as a guideline for computing the optimal control law.

B. Completely Observable Controller

It is known (Fleming [26, 28]) that Markovian optimization problems can be treated well using dynamic programming technique. In the Markovian problem the controller is allowed complete information about the current state. The system is described by the Ito stochastic differential equation in R^n-space

$$dx(t) = f(t,x(t),u(t,x(t)))\,dt + g(t,x(t))\,dw(t), \qquad (5.1)$$

OPTIMAL CONTROL OF STOCHASTIC SYSTEMS 37

$t \in I = [0, T]$, with given initial data $x(0) = x_0$; w, as before, is the Wiener process in R^m, f and g are of appropriate dimensions. The problem is: Find a control in an admissible class $\mathcal{U}$ that minimizes the average cost functional

$$J(u) = E \int_0^T L(t, x(t), u(t, x(t))) \, dt, \qquad (5.2)$$

$E(.)$ denoting expected value, where x is the response to the control u. Fleming introduced the following assumptions:

(F1) *The admissible controls consist of functions* $u : Q_T \to U$, *where* $Q_T \equiv I \times R^n$, U *a fixed compact convex subset of some p-dimensional space* R^p *and for each* $0 \leq T' < T$, u *is u.h.c.* (α) *in t (uniformly Hölder continuous with exponent $0 < \alpha < 1$) and u.l.c. in ξ on $Q_{T'}$ such that*

$$|u(t', \xi) - u(t, \xi)| \leq M|t' - t|^\alpha,$$
$$|u(t, \xi') - u(t, \xi)| \leq M|\xi' - \xi|$$

for some $M \in (0, \infty)$ and for all $\xi, \xi' \in R^n$ and $0 \leq t, t' \leq T'$.

Let $\mathcal{U}$ denote the set of all admissible controls.

(F2) $f : Q_T \times U \to R^n$ *and* $g : Q_T \to R^{n \times m}$ *(the space of $n \times m$ matrices with real-valued elements) are u.h.c. in t and satisfy, for suitable positive constants M_1, M_2,*

$$|f(t, \xi, y)| \leq M_1, \qquad |g(t, \xi)| \leq M_1,$$
$$|f(t, \xi', y) - f(t, \xi, y)| \leq M_2|\xi' - \xi|, \qquad |g(t, \xi') - g(t, \xi)| \leq M_2|\xi' - \xi|$$

for every $\xi, \xi' \in R^n$, $0 \leq t, t' \leq T$, and $y \in U$.

(F3) $L : Q_T \times U \to R$ *is u.h.c. on $Q_T \times U$ and satisfies for suitable positive constant M_3, $|L(t, \xi, y)| \leq M_3$, $|L(t, \xi', y) - L(t, \xi, y)| \leq M_3|\xi' - \xi|$ for every $\xi, \xi' \in R^n$, $0 \leq t, t' \leq T$, and $y \in U$.*

Under the assumptions (F1) and (F2) the system (5.1) is known to have a solution x which is a continuous Markov process satisfying

$$E \int_0^T |x(t)|^2 \, dt < \infty.$$

The equality in (5.1) holds with probability one.

Exit Times Let B be a compact subset of R^n with the boundary $\partial B \in C^{2+\alpha}$, $0 < \alpha < 1$ (with Hölder continuous second partial derivatives). Given initial data $x(0) = x_0 \in B$, let us stop the process x at the first time $\tau < T$ when $x(\tau) \in \partial B$; if $x(t) \in B \setminus \partial B$ for $t \in [0, T]$, then set $\tau = T$. The random variable τ is called the exit time from the cylinder $Q = I \times B$ and it is known to be nonanticipative in the sense that the set $\{\omega : \tau < t\}$ is $B_t \equiv \sigma\{w(s) : s \leq t\}$-measurable. For the optimization problem as stated above the condition

(F3) is rather too restrictive. However, for the same problem corresponding to the stopped process this condition is sufficiently general. Here we will consider the stopped process. Let $s \in I$ be the initial time, $x(s) = \xi \in B$ the initial state, and $u \in \mathcal{U}$ and x the corresponding response of the system (5.1). Define

$$\phi^u(s, \xi) = E\left\{\int_s^\tau L(t, x(t) : u(t, x(t))) \, dt \,|\, x(s) = \xi\right\} \qquad (5.3)$$

to be the expected cost corresponding to the control u and initial state (s, ξ). The problem is to minimize ϕ^u on $\mathcal{U}$. Define

$$a(s, \xi) = \tfrac{1}{2} g(s, \xi) g'(s, \xi) \qquad (' \text{ denotes matrix transpose})$$

and assume

(F4) a is positive definite over $Q = I \times B$ and there exists $c > 0$ such that $\sum_{i,j} a_{i,j} \lambda_i \lambda_j \geq c|\lambda|^2$ for all $\lambda \in R^n$ uniformly in Q.

For each control $u \in \mathcal{U}$, under the assumptions (F2), (F3), and (F4) the function ϕ^u satisfies [67] the linear parabolic partial differential equation with the first boundary condition:

$$\Lambda(\phi^u) + f^u \cdot \phi^u_\xi + L^u = 0, \qquad (s, \xi) \in Q,$$
$$\phi^u = 0, \qquad (s, \xi) \in \Sigma \equiv I \times \partial B \cup \{T\} \times B, \qquad (5.4)$$

where

$$\Lambda(\psi) = \psi_t + \sum_{i,j=1}^n a_{ij} \psi_{\xi_i, \xi_j},$$

$$f^u \cdot \psi_\xi = \sum_{i=1}^n f_i(s, \xi; u(s, \xi)) \psi_{\xi_i},$$

$$L^u = L(s, \xi, u(s, \xi)),$$

and $\phi^u \in \mathcal{F}_0$, where the class $\mathcal{F}_0$ consists of real-valued functions $\{\psi\}$ defined on Q with the properties

  (i) ψ and ψ_ξ are Holder continuous on Q,
  (ii) $\psi_s, \psi_{\xi_i, \xi_j}, i, j = 1, 2, \ldots n$, are continuous on $Q \backslash \{T\} \times \partial B$ and $\psi_s, \psi_{\xi_i, \xi_j} \in L_2(Q, R)$, and
  (iii) $\psi(s, \xi) = 0$ for $(s, \xi) \in \Sigma$.

Fleming then shows that the function ψ defined as

$$\psi(s, \xi) = \inf_{u \in \mathcal{U}} \phi^u(s, \xi), \qquad (s, \xi) \in Q, \qquad (5.5)$$

satisfies a quasi-linear parabolic differential equation.

For each $(s, \xi) \in Q$ and $z \in R^n$ define the function

$$H(s, \xi, z) = \min_{v \in U} [L(s, \xi, v) + f(s, \xi, v) \cdot z], \tag{5.6}$$

where $f \cdot z = \sum_i f_i z_i$.

Theorem 5.1 *The function ψ defined by (5.5) belongs to the class $\mathcal{F}_0$ and satisfies the quasi-linear parabolic equation*

$$\begin{aligned}\Lambda(\psi) + H(s, x, \psi_x) &= 0, \quad (s, x) \in [0, T) \times R^n, \\ \psi(T, x) &= 0, \quad x \in R^n.\end{aligned} \tag{5.7}$$

The proof of this result is given in Fleming [26, Theorem 2.1] and is mainly based on an existence and uniqueness theorem for the solution of the system (5.7) and the maximum principle for linear parabolic differential equations of the type (5.4).

The significance of this result is that (5.6) and (5.7) provide a set of necessary conditions and that if an optimal control exists in the class $\mathcal{U}$, then this control can be computed in three steps:

Step 1 Construct a function $y = y(s, \xi, z)$ on $Q \times R^n$ from the relation (5.6), so that

$$H(s, \xi, z) = L(s, \xi, y(s, \xi, z)) + f(s, \xi, y(s, \xi, z)) \cdot z.$$

Step 2 Solve the quasi-linear parabolic equation (5.7) in Q for $\psi \in \mathcal{F}_0$.

Step 3 Set $u^*(s, \xi) = y(s, \xi, \psi_\xi(s, \xi))$.

The question of existence of optimal control in the class $\mathcal{U}$, however, remains critical. For example, it may be easily verified that if $L \equiv 1$ ($\phi^{u(s,\xi)} = E\{\tau | x(s) = \xi\}$), f linear in u and U a compact hypercube in R^p-then an optimal control may not exist in the class $\mathcal{U}$; in fact the minimizing control may well be discontinuous. This is well known for the Pontryagin problem. However, there are sufficient conditions (Fleming [26, Theorem 2.2, p. 261]), namely,

(i) f, L of class C^2,
(ii) f linear in control, and
(iii) $L_{u_i u_j} \theta_i \theta_j \geq \beta |\theta|^2$ for all $\theta \in R^p$, $\beta > 0$, uniformly on $Q \times U$,

under which there exists a control $u^* \in \mathcal{U}$ such that $\psi(s, \xi) = \phi^{u^*}(s, \xi)$ and $L^u + f^u \cdot \psi_\xi$ attains its minimum at $u = u^*(s, \xi)$. However, if measurable controls are admitted then the above conditions may be relaxed, as observed in Theorems 3.2, 3.9, and 3.10.

It is known that application of dynamic programming techniques to solve Lagrange problems consisting of a deterministic dynamical system $dx(t) = f(t, x(t), u(t)) dt$ and the cost functional $J(u) = \int_0^T L(t, x(t), u(t)) dt$

gives rise to similar conditions as (5.6) and (5.7) with the diffusion term missing from the operator Λ. Clearly, the corresponding equation is a first-order hyperbolic equation. Thus the above result can be considered as an extension of the classical Lagrange problem.

C. Partially Observable Controller with Past Information

In many practical situations the controller is required to decide on its control action based on the information (data) available from the instrumentation channel, which may be noisy as well. This problem was first treated by Wonham [72], who presented a separation principle (see Section III.C, Theorem 3.1), thereby generalizing a similar result due to Joseph and Tou [40] and Potter [55] for linear stochastic regulator problems. Let us consider a stochastic system in which the plant to be controlled is described by the stochastic linear differential equation

$$S_1 \quad \begin{cases} dx(t) = [A(t)x(t) + b(t, u(t))]\, dt + B(t)\, dw_1(t), \\ x(0) = x_0 \end{cases}$$

and the measurement channel is described by

$$C \quad \begin{cases} dy(t) = F(t)x(t)\, dt + G(t)\, dw_2(t), \\ y(0) = 0. \end{cases}$$

The problem is to minimize the cost function

$$J(u) = E \int_0^T L(t, x(t), u(t))\, dt$$

choosing $u(t) = \psi(t, \pi_t y)$, where $\psi \in \Psi$, the class of nonanticipative Volterra functionals with values in a compact convex set U as described in Section III. As we have seen in Section III, due to the separation principle (Theorem 3.1) this problem can be reduced to one of standard filtering and Markovian control. It can be shown [72, 74] that $\hat{x}(t) \equiv E\{x(t)|Y_t\}$ satisfies the following stochastic differential equation:

$$d\hat{x}(t) = [A(t)\hat{x}(t) + b(t, \hat{u}(t, \hat{x}(t)))]\, dt + QF'(GG')^{-1/2}\, d\hat{w}, \quad (5.8)$$
$$\hat{x}(0) = Ex(0)$$

for a given $\hat{u} \in \hat{\Psi}$ (Section III.C), where $\hat{w}$ is a standard Wiener process given by

$$d\hat{w} = (GG')^{-1/2}(dy - F\hat{x}\, dt). \quad (5.9)$$

Further, the conditional covariance matrix Q defined as $Q(t) = E\{(x(t) - \hat{x}(t))(x(t) - \hat{x}(t))'|Y_t\}$ satisfies the Kalman matrix Riccati equation [72]

$$dQ/dt = AQ + QA' + BB' - Q(F'(GG')^{-1}F)Q, \quad t \in I,$$
$$Q(0) = Q_0 = E(x_0 x_0'). \quad (5.10)$$

It is also known that the conditional distribution of $x(t)$ given the σ-algebra Y_t is Gaussian with mean $\hat{x}(t)$ and covariance matrix $Q(t)$. Thus the original optimization problem consisting of the dynamical constraints S_1, and C and the cost function J, now reduces to the (Markovian) optimization problem consisting of the dynamical constraint (5.8) and the cost function

$$J(\hat{u}) = E \int_0^T \hat{L}(t, \hat{x}(t), \hat{u}(t, \hat{x}(t))) \, dt \qquad (5.11)$$

with $\hat{u} \in \hat{\Psi}$, where

$$\hat{L}(t, \xi, v) = \int_{R^n} L(t, x, v) \zeta(x, t, \xi) \, dx, \xi \in R^n, v \in U,$$

and

$$\zeta(x, t, \xi) = (2\pi)^{-n/2} [\det Q(t)]^{-1/2} \exp[-\tfrac{1}{2}(x - \xi)' Q(t)^{-1}(x - \xi)].$$

At this point as discussed before, by application of dynamic programming technique, to (5.8) and (5.11) we obtain the optimal control laws. The significance of the separation theorem is best illustrated by the block diagram of Fig. 1, the arrows indicating the direction of information flow. In order to find the optimal control law one follows the following steps:

Step 1 Solve the Kalman matrix Riccati equation

$$dQ/dt = AQ + QA' + BB' - Q(F'(GG')^{-1}F)Q, \qquad t \in I,$$
$$Q(0) = Q_0 \equiv E(x_0 x_0').$$

Step 2 Compute $\hat{L}(t, \xi, v)$,

$$\hat{L}(t, \xi, v) = \int_{R^n} L(t, x, v) \zeta(x, t, \xi) \, dx,$$

where

$$\zeta(x, t, \xi) = (2\pi)^{-n/2} [\det Q(t)]^{-1/2} \exp\{-\tfrac{1}{2}(x - \xi)' Q(t)^{-1}(x - \xi)\}$$

is the conditional probability density of $x(t)$ given $\hat{x}(t) = \xi$ at time $t \in I$.

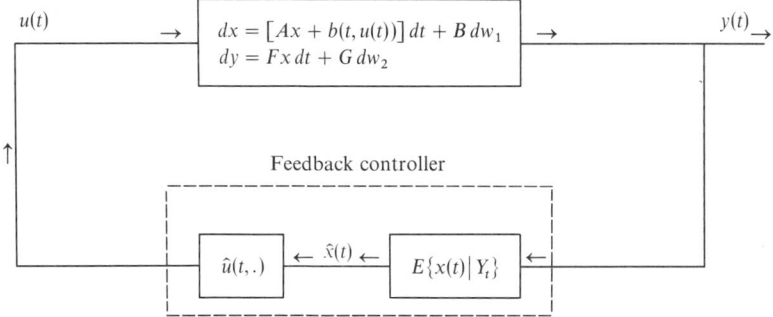

Fig. 1 Separation theorem in block diagram.

Step 3 Define $\lambda(t,\xi,p,v) = b(t,v)'p + \hat{L}(t,\xi,v)$ and minimize $\lambda(t,\xi,p,v)$ with respect to v over U giving $v = \mu(t,\xi,p)$ for each $(t,\xi,p) \in I \times R^n \times \{z : |z| < \beta\}$, for a suitable $0 < \beta < \infty$. That is,

$$\min_{v \in U} \lambda(t,\xi,p,v) = \lambda(t,\xi,p,\mu(t,\xi,p)).$$

The existence and uniqueness of the function μ (with values in U) is known (Fleming [26, Lemma 2.1], Wonham [72, Lemma 5.3]).

Step 4 Solve the quasilinear parabolic equation with Cauchy condition

$$\phi_t(t,\xi) + \tfrac{1}{2}\operatorname{tr}\{\hat{C}(t)\phi_{\xi\xi}(t,\xi)\hat{C}(t)\} + \xi' A'(t)\phi_\xi(t,\xi)$$
$$+ \lambda(t,\xi,\phi_\xi(t,\xi),\mu(t,\xi,\phi_\xi(t,\xi))) = 0, \quad (t,\xi) \in I \times R^n,$$
$$\phi(T,\xi) = 0, \quad \xi \in R^n,$$

where $\hat{C} \equiv QF'(GG')^{-1/2}$.

Using results of Ladyzenskaya et al. [49, Theorem 8.1, p. 495] on the existence and uniqueness of solutions of such quasi-linear parabolic equations, Wonham established the existence of smooth solution of the above Cauchy problem.

Step 5 The optimal control law is then given by $\hat{u}^0(t,\xi) = \mu(t,\xi,\phi_\xi(t,\xi))$. That this is the optimal control law follows from the fact that $\hat{u}^0$ satisfies the optimality criterion [74, Lemma 5.2, p. 189] as given by Wonham.

Step 6 For complete synthesis of the optimal feedback controller, the filter equation

$$d\hat{x} = [A\hat{x} + b(t,\hat{u}^0(t,\hat{x})) - KF\hat{x}]\,dt + K\,dy, \quad t \in I,$$
$$\hat{x}(0) = Ex_0,$$

where $K = QF'(GG')^{-1}$, is to be solved in real time with the evolution of the information $\{y(t) : t \in I\}$. This may be implemented by real time simulation of the filter equation on analog computer.

As stated in Section III, Fleming [25] proved a revised version of the separation principle under weaker assumptions on the cost integrand L and the control constraint set U. However, as of now it is not known whether a separation principle can be proved for nonlinear stochastic systems.

D. Partially Observable Controller with Only Current Information

Although no necessary conditions of optimality appear to have been proved for nonlinear systems involving a plant (to be controlled) and a measurement channel, as in Wonham, recently Fleming formulated [30] an interesting class of control problems in which the controller is allowed to observe, and base its control on the current values of part of the states. The plant to be controlled is described by the stochastic differential equation

$$dx = f(t, x, u(t, \tilde{x})) dt + g(t, x, u(t, \tilde{x})) dw, \quad t \in I = [0, T], \quad x(0) = x_0, \quad (5.12)$$

where $x = (x_1, x_2, \ldots, x_n) \in R^n$—is the dynamic state, $\tilde{x} = (x_1, x_2, \ldots, x_l) \in R^l$, $0 \leq l \leq n$, is its first l components, $\tilde{\tilde{x}} = (x_{l+1}, x_{l+2}, \ldots, x_n) \in R^{n-l}$ is the last $(n-l)$ components, and the cost functional is given by

$$J(u) = E \int_0^\tau L(t, x(t), u(t, \tilde{x}(t))) dt, \quad (5.13)$$

with τ being the first exit time from the cylinder $Q = (0, T) \times B$ as defined in Section III and $B \subset R^n$ an open set with compact closure. The problem is to find a control u in the class of admissible controls $\mathscr{U}$, to be defined shortly, so that $J(u)$ is a minimum. It is clear that Fleming's problem is a generalization of Wonham's problem subject to the condition that in the latter, the controller is allowed to observe only the current values of the channel output y. For this problem Fleming [30] has presented necessary conditions of optimality under two different situations: (i) both the drift and diffusion matrices f, g contain controls and the admissible class of controls consists of functions $\{u(t, \tilde{x})\}$ that are bounded and Lipschitz with values in a closed convex set U; and (ii) the diffusion matrix g is independent of control and the class of admissible controls consists of functions $\{u(t, \tilde{x})\}$ that are bounded and measurable with values in a closed convex set U. For convenience we will denote the class of smooth controls by $\mathscr{U}_s$ and the class of measurable controls by $\mathscr{U}_m$. It is known that the above optimality problem is equivalent to the optimization problem of the following first boundary value problem:

$$A^u \psi + L^u = 0 \quad \text{in} \quad Q \equiv (0, T) \times B,$$
$$\psi = 0 \quad \text{on} \quad \Sigma \equiv \{[0, T] \times \partial B\} \cup \{\{T\} \times B\}, \quad (5.14)$$
$$J(u) = \int_B \psi(0, x) d\pi_0 = \min,$$

where

$$A^u \psi \equiv \psi_t + a^u \cdot \psi_{xx} + f^u \cdot \psi_x,$$
$$a^u(t, x) \equiv a(t, x, u(t, \tilde{x})) \equiv \tfrac{1}{2} g(t, x, u(t, \tilde{x},)) g'(t, x, u(t, \tilde{x})),$$
$$f^u(t, x) \equiv f(t, x, u(t, \tilde{x})),$$
$$L^u(t, x) \equiv L(t, x, u(t, \tilde{x})),$$

and π_0 is the initial probability measure with support B.

Necessary Conditions with Controls $\mathscr{U}_s$ For the existence and uniqueness of solutions of the first boundary problems (5.14) and the corresponding adjoint problem in the weak form as discussed below, it is essential to introduce the following assumptions:

(i) f, g, L are of class C^2;
(ii) if $U \subset R^p$ is any compact set, then f, g, f_x, g_x are bounded on $[0, T] \times B \times U$, $B \subset R^n$ open and bounded;
(iii) there exists a constant $\beta > 0$ such that $a_{ij}(t, x, u)\lambda_i\lambda_j \geq \beta |\lambda|^2$ for all $\lambda \in R^n$ independently of $(t, x, u) \in [0, T] \times B \times U$.

The assumptions (i) and (ii) will be relaxed in case of measurable controls. However, condition (iii) is essential to ensure uniform parabolicity of the first boundary problem (5.14) and consequent uniqueness of its solution.

In case the class $\mathscr{U}_s$ is chosen for admissible controls, all the coefficients of the operator A^u may depend on the control. Under these conditions it is known that the first boundary problem in (5.14) has a unique solution in the class $\mathscr{F}_0$ (Section V.B). Further, if π_0 has a smooth density q_0, then the adjoint boundary problem

$$\begin{aligned}(A^u)^*q &= 0, & (t, x) &\in Q, \\ q(t, x) &= 0, & (t, x) &\in [0, T] \times \partial B, \\ q(0, x) &= q_0(x), & x &\in B,\end{aligned} \quad (5.15)$$

has a unique weak solution $q \in L_1(\bar{Q}, R)$ in the sense that

$$\int_Q (A^u\psi, q)\, dt\, dx = -\int_B \psi(0, x) q_0(x)\, dx \quad (5.15')$$

for every $\psi \in \mathscr{F}_0$ for which $A^u\psi$ is bounded. In other words, $q(t,.)$ is the probability density of the random vector $x^0(t)$, where x^0 is the response process x stopped at Σ. Furthermore, if q_0 is assumed to be of class C^1 with bounded first partials, then the weak solution q satisfies

(i) $q \in L_1(Q, R)$,
(ii) for $\delta > 0$, q is Hölder continuous on $\bar{Q}_\delta$, $Q_\delta \equiv (\delta, T) \times B$ and $q_x \in L_2(Q_\delta, R^n)$,
(iii) $q(t, x) = 0$ for $0 < t \leq T$ and $x \in \partial B$, and
(iv) $q(t, x) > 0$ for $0 < t \leq T$ and $x \in B$.

For convenience of description we will denote this class of functions by $\mathscr{F}_1^*$. Let Φ be any continuous real-valued function defined on $(0, T) \times \bar{B}$ ($\bar{B} =$ closure of B) with values $\Phi(t, x) = \Phi(t, \tilde{x}, \tilde{\tilde{x}})$ and define

$$\hat{\Phi}(t, \tilde{x}) \equiv \int_{R^{n-1}} \Phi(t, \tilde{x}, \tilde{\tilde{x}}) q_c(t, \tilde{\tilde{x}} | \tilde{x})\, d\tilde{\tilde{x}}$$

where

$$q_c(t,\tilde{\tilde{x}}|\tilde{x}) = \frac{q(t,\tilde{x},\tilde{\tilde{x}})}{\int_{R^{n-l}} q(t,\tilde{x},\tilde{\tilde{x}})\,d\tilde{\tilde{x}}} \equiv \frac{q(t,\tilde{x},\tilde{\tilde{x}})}{\hat{q}(t,\tilde{x})}$$

is the conditional probability density of $\tilde{\tilde{x}}(t)$ given $\tilde{x}(t)$. Let us assume that $u^0 = u^0(t,\tilde{x}) \in \mathcal{U}_s$ is the optimal control (existence assumed; see Section III). Let ψ^0 be the solution of the first boundary problem in (5.14) corresponding to the control u^0. Define, for each $v \in U$,

$$H(t,x,v) \equiv a^v\psi^0_{xx} + f^v \cdot \psi^0_x + L^v \tag{5.16}$$

and

$$\hat{H}(t,\tilde{x},v) = \int_{R^{n-l}} H(t,\tilde{x},\tilde{\tilde{x}},v) q^0_c(t,\tilde{\tilde{x}}|\tilde{x})\,d\tilde{\tilde{x}}, \tag{5.17}$$

where q^0 is the weak solution of the adjoint boundary problem (5.15) corresponding to the control u^0. With this preparation we are now in a position to state the necessary conditions of optimality as given by Fleming [30, Theorem 1, p. 201].

Theorem 5.2 *If $u^0 \in \mathcal{U}_s$ is optimal, minimizing the cost function $J(u)$, then for every $(t,\tilde{x}) \in \tilde{Q}$, $\tilde{Q} \equiv \{(t,\tilde{x}) \in (0,T) \times R^l : (t,\tilde{x},\tilde{\tilde{x}}) \in Q\}$,*

$$\hat{H}_u(t,\tilde{x},u^0(t,\tilde{x})) \cdot v \geq 0 \quad \text{for all} \quad v \in U(u^0(t,\tilde{x})), \tag{5.18}$$

where

$$U(v) \equiv \{z \in R^p : v + \alpha z \in U, 0 \leq \alpha \leq h(z)\},$$

$h(z) > 0$, *is the convex cone, called the contingent to U at v.*

The proof of this result easily follows from the relations (5.14), (5.15'), and an application of the technique of first variation in the calculus of variations.

(i) In case $U \equiv R^p$ (no control constraints), the above necessary condition reduces to

$$\hat{H}_u(t,\tilde{x},u^0(t,\tilde{x})) = 0 \quad \text{for every} \quad (t,\tilde{x}) \in \tilde{Q}. \tag{5.19}$$

(ii) In case L is convex in u and f and a are linear in u, then the above necessary condition implies that $\hat{H}(t,\tilde{x},v)$ attains its minimum at $v = u^0(t,\tilde{x})$.

Thus in order to determine the optimal control law u^0, it is required to solve the initial boundary problems (5.14) and (5.15') and the necessary condition (5.18) simultaneously. In certain special situations [8] it is possible to determine the form of the control law from the relation (5.18) as a function of $(t,\tilde{x})$ and functional of ψ and q. By substituting this functional relation in (5.14), (5.15), or (5.15') one obtains two coupled systems of nonlinear integropartial differential equations with first boundary conditions [8]. The optimal control is to be determined by solving these equations, which is by no means trivial. Thus the problem of computation of partial observer–controller is far more complex than that for the complete observer–controller, for the latter of which the equation of Bellman's dynamic programming (5.7) is to be solved.

Necessary Conditions with Controls $\mathcal{U}_m \subset L_\infty(\tilde{Q}, R^p)$ In case the diffusion matrix a is independent of the control variable, measurable controls can be admitted. The assumptions of the previous section are replaced by

(i) f, L are measurable on Q for each $u \in U$ and continuous on U for almost all $(t, x) \in Q$,

(ii) $a(t, x) = \frac{1}{2}(gg')(t, x)$ is of class C^2 and $a_{ij}(t, x)\lambda_i\lambda_j \geq \beta|\lambda|^2$ for some $\beta > 0$, for all $\lambda \in R^n$ uniformly on Q.

Let $\mathcal{F}_1$ denote the class of functions $\mathcal{F}_0$ (Section V.B) with the condition (ii) replaced by (ii'): $\psi_s, \psi_{x_i,x_j} \in L_2(Q, R), i, j = 1, 2, \ldots n$, only. It is known (Fleming [30, p. 202]) that under the above assumptions the boundary problem (5.14) has a unique solution in the class $\mathcal{F}_1$ for every control $u \in \mathcal{U}_m$. Similarly, the adjoint boundary problem (5.15) has a unique weak solution in the class $\mathcal{F}_1^*$. Under the above assumptions Fleming [30, Theorem 2, p. 202] presented a necessary condition of optimality analogous to that of Pontryagin.

Theorem 5.3 *Suppose there exists a control $u^0 \in \mathcal{U}_m$ that minimizes the cost functional $J(u)$ over $\mathcal{U}_m$ and let $\psi^0 \in \mathcal{F}_1$ and $q^0 \in \mathcal{F}_1^*$ be the corresponding solutions of (5.14) and (5.15'), respectively. Define*

$$H(t, x, v) = f^u \psi_x^0 + L^v$$

and

$$\hat{H}(t, \tilde{x}, v) = \int_{R^{n-l}} H(t, \tilde{x}, \tilde{\tilde{x}}, v) q_c^0(t, \tilde{\tilde{x}} | \tilde{x}) d\tilde{\tilde{x}},$$

where q_c^0 is the conditional density corresponding to the control u^0. Then

$$\hat{H}(t, \tilde{x}, u^0(t, \tilde{x})) \leq \hat{H}(t, \tilde{x}, v) \tag{5.20}$$

for all $v \in U$ and for almost all $(t, \tilde{x}) \in \tilde{Q}$.

Using Sobolev's lemma and the relations (5.14) and (5.15'), the proof is given by establishing a contradiction to the reverse inequality considered on $\tilde{Q}$ (except on a set of measure zero) for a countable dense subset of U. According to Theorem 5.3, it is clear that for $l = n$, that is, for the completely observed case, $\hat{H}(t, \tilde{x}, v) = H(t, x, v)$ and the minimum condition (5.20)

OPTIMAL CONTROL OF STOCHASTIC SYSTEMS

reduces to

$$H(t, x, u^0(t, x)) \leq H(t, x, v), \quad v \in U, \quad (t, x) \in Q.$$

The optimal control is determined by solving the equation of dynamic programming

$$\Lambda(\psi^0) + \min_{v \in U} H(t, x, v) = 0 \quad \text{in} \quad Q,$$

$$\psi^0 = 0 \quad \text{in} \quad \Sigma,$$

as discussed before. In this case the adjoint problem is not involved.

At the other extreme, for $l = 0$, that is, in the absence of any information,

$$\hat{H}(t, \tilde{x}, v) = \hat{H}(t, v) \equiv \int_{R^n} H(t, x, v) q^0(t, x) \, dx$$

and the corresponding minimum condition is given by

$$\hat{H}(t, u^0(t)) \leq \hat{H}(t, v), \quad v \in U, \quad t \in [0, T].$$

The control is now only a function of t and a functional of ψ_x^0 and q^0, that is,

$$u^0(t) = \mu(t, \psi_x^0, q^0)$$

where

$$\mu : I \times C(B, R^n) \times C(B, R) \to U \subset R^p$$

and minimizes the function $\hat{H}(t, v)$ over U. For the optimal control we must now solve the following first boundary problems involving functional partial differential equations:

$$\Lambda(\psi^0) + \hat{H}(t, \mu(t, \psi_x^0, q^0)) = 0 \quad \text{in} \quad Q,$$
$$\psi^0 = 0 \quad \text{in} \quad \Sigma, \tag{5.21}$$

and

$$\int_Q [\Lambda(\psi^0) + f(t, x, \mu(t, \psi_x^0, q^0)) \cdot \psi_x^0] q^0 \, dt \, dx = -\int_B \psi^0(0, x) q_0(x) \, dx \tag{5.22}$$

for $(\psi^0, q^0) \in \mathscr{F}_1 \times \mathscr{F}_1^*$. For all intermediate values of $0 < l < n$, we have to solve similarly complex functional partial differential equations with first boundary conditions. If $\mu : \tilde{Q} \times C(\tilde{B}, R^n) \times C(\tilde{B}, R) \to U$ minimizes $\hat{H}(t, \tilde{x}, v)$ over U, then the optimal control law is given by $u^0(t, \tilde{x}) = \mu(t, \tilde{x}, \psi_x^0, q_c^0)$. In case of time optimal control [e.g., $\max\{E(\tau) : u \in U, \tau =$ first time $\leq T$, $x(\tau) \in \partial B\}$] with f linear in u, it is known (Ahmed and Teo [8, Theorem 1, p. 74]) that the bang-bang principle holds. In fact the smoothness of control depends on that of the cost integrand L considered as a function of u. In fact if L is strictly convex in the control variable $u \in U$ with $L_{u_i u_j}(t, x, u) \beta_i \beta_j \geq$

$\gamma|\beta|^2$, $(t, x, u) \in Q \times U$, then the optimal control $u^0 = u^0(t, \tilde{x})$ is Hölder continuous on any compact subset of $\tilde{Q}$ [30, Theorem 4, p. 205]. In all the above necessary conditions three major steps are to be followed to compute the optimal control law:

Step 1 Determine $\mu = \mu(t, \tilde{x}, \psi_x^0, q_c^0)$ so that

$$\hat{H}(t, \tilde{x}, \mu(t, \tilde{x}, \psi_x^0, q_c^0)) = \min_{v \in U} \hat{H}(t, \tilde{x}, v).$$

Step 2 Solve the following functional partial differential equations for ψ^0 and q^0:

$$\Lambda(\psi^0) + \hat{H}(t, \tilde{x}, \mu(t, \tilde{x}, \psi_x^0, q_c^0)) = 0 \quad \text{in} \quad Q,$$
$$\psi^0 = 0 \quad \text{in} \quad \Sigma,$$

and

$$\int_Q [\Lambda(\psi^0) + f(t, x, \mu(t, \tilde{x}, \psi_x^0, q_c^0)) \cdot \psi_x^0] q^0 \, dt \, dx = -\int_Q \psi^0(0, x) q_0(x) \, dx$$
$$q^0(0, x) = q_0(x).$$

Step 3 Set $u^0(t, \tilde{x}) = \mu(t, \tilde{x}, \psi_x^0, q_c^0)$.

It is clear that the Steps (1) and (2) are by no means trivial. An interesting problem would be to develop a computational algorithm for solving these two-point boundary value problems.

E. *Optimal Policy Involving Controls and Parameters*

In problems of engineering design it is often essential [9, 66] to choose component values from an admissible set (called the parameter space), in addition to choosing controls from an admissible class, so that the final product has the best performance in some appropriate sense. This problem can be formulated as follows: Consider the stochastic system

$$S \begin{cases} dx(t) = f(t, x(t), \sigma, u(t, \tilde{x}(t))) \, dt + g(t, x(t), \sigma) \, dw(t), & t \in [0, T] \equiv I, \\ x(0) = x_0 & (\pi_0 \text{ denotes the initial probability measure,} \\ \sigma \in \Gamma & ((\Gamma \text{ denotes the parameter space}), \\ u \in \mathcal{U} & (\mathcal{U} \text{ denotes the class of admissible controls}), \\ w = \text{Wiener (vector) process as before,} \end{cases}$$

with the performance functional

$$J(\sigma, u) = E\left\{\int_0^T L(t, x(t), \sigma, u(t, \tilde{x}(t))) \, dt + \eta(x(T))\right\},$$

where $\tilde{x}$ again denotes the first l components of the state vector x. The problem is to find an optimal policy $(\sigma^0, u^0) \in \Gamma \times \mathcal{U}$ that minimizes the functional $J(\sigma, u)$. The policy space is given by the cartesian product $\Gamma \times \mathcal{U}$, where Γ is a closed convex subset of a finite-dimensional parameter space R^q and $\mathcal{U}$ is the class of measurable functions on $[0, T] \times R^l$ with values in a compact convex subset U of R^p. Recently, Ahmed and Teo [6, Theorem 6.3] have presented necessary conditions for optimality for the above system. The basic assumptions utilized in their paper are:

(A*1) $a_{ij}(t, x, \sigma) \equiv \frac{1}{2}(gg')_{ij}(t, x, \sigma)$ $(i, j = 1, 2, \ldots n)$ are measurable on I for each $(x, \sigma) \in R^n \times R^q$ and continuous on $R^n \times R^q$ for almost all $t \in I$ and are all jointly continuous functions of their arguments. Further, there exist constants $\alpha_l, \alpha_u > 0$ such that $\alpha_l |z|^2 \leq a_{ij}(t, x, \sigma) z_i z_j \leq \alpha_u |z|^2$ for all $z \in R^n$ uniformly on $I \times R^n \times \Gamma$.

(A*2) $\partial a_{ij}/\partial x_j$ $(i, j = 1, 2, \ldots n)$ are bounded measurable on I for each $(x, \sigma) \in R^n \times \Gamma$ and continuous bounded on $R^n \times \Gamma$ for almost all $t \in I$, and f_i $(i = 1, 2, \ldots n)$ are bounded measurable on I for each $(x, \sigma, v) \in R^n \times \Gamma \times U$ and continuous bounded on $R^n \times \Gamma \times U$ for almost all $t \in I$.

(A*3) L is measurable on I for each $(x, \sigma, v) \in R^n \times \Gamma \times U$ and continuous on $R^n \times \Gamma \times U$ for almost all $t \in I$, and $L(., ., \sigma, u) \in L^{p'}(I, L^2(R^n, R))$ for every $(\sigma, u) \in \Gamma \times U$, where $p' \in (1, 2]$.

(A*4) $\eta : R^n \to R$ is continuous, so that $\eta \in L^2(R^n, R)$.

It is known [6, Theorem 3.1] that the system S, as described above, has a unique solution $x(\sigma, u)$ over I for each $(\sigma, u) \in \Gamma \times \mathcal{U}$ and that $\{x(\sigma, u)(t) : t \in I\}$ is a Markov process. Using this fact and the assumption (A*2) it is easily shown that the above optimization problem is equivalent to

$$S' \begin{cases} \mathscr{L}(\sigma, u)\phi(\sigma, u)(t, x) + L(t, x, \sigma, u(t, \tilde{x})) = 0, & (t, x) \in I \times R^n, \\ \phi(\sigma, u)(T, x) = \eta(x), x \in R^n, & (\sigma, u) \in \Gamma \times \mathcal{U}; \end{cases}$$

$$\min_{(\sigma, u) \in \Gamma \times \mathcal{U}} J(\sigma, u) = \min_{(\sigma, u) \in \Gamma \times \mathcal{U}} \int_{R^n} \phi(\sigma, u)(0, x) \, d\pi_0(x),$$

where

$$\mathscr{L}(\sigma, u)\psi \equiv \psi_t + (a_{ij}\psi_{x_i})_{x_j} - a_i \psi_{x_i},$$

$$a_i \equiv \sum_{j=1}^{n} (\partial a_{ij}/\partial x_j) - f_i.$$

Clearly, this is a Cauchy problem, as expected for fixed time problems [67, Theorem 1]. It has been shown [6, Theorem 4.2] that for each admissible policy $(\sigma, u) \in \Gamma \times \mathcal{U}$, the Cauchy problem S', under the assumptions

(A*1)–(A*4), has a unique weak solution $\phi(\sigma, u) \in L^\infty(I, L^2(R^n, R)) \cap L^2(I, W'(R^n, R))$ in the sense that

$$\int_Q [-\phi(\sigma, u)\psi_t - a_{ij}\phi_{x_i}\psi_{x_j} - a_i\phi_{x_i}\psi + L\psi]\,dt\,dx = 0$$

for every $\psi \in C'_0(R^n, R)$ and

$$\lim_{t \to T} \int_{R^n} \phi(\sigma, u)(t, x)z(x)\,dx = \int_{R^n} \eta(x).z(x)\,dx$$

for all $z \in C'_0(R^n, R)$, where W' is the completion of C^∞ functions with compact support in the norm

$$\|z\|_2 + \|z_x\|_2 \equiv \left(\int_{R^n} |z(x)|^2\,dx + \int_{R^n} \sum_{i=1}^{n} |z_{x_i}|^2\,dx\right)^{1/2}$$

and C'_0 is the space of once continuously differentiable functions with compact support. This result remains valid also for the adjoint Cauchy problem:

AS' $\begin{cases} \mathscr{L}^*(\sigma, u)q = 0 & \text{for } (t, x) \in Q \equiv (0, T] \times R^n, \\ q(0, x) = q_0(x) & \text{for } x \in R^n, \end{cases}$

where $\mathscr{L}^*(\sigma, u)\phi = -\phi_t + \{a_{ij}\phi_{x_j} + a_i\phi\}_{x_i}$, and for any measurable $E \subset R^n$, $\pi_0(E) = \int_E q_0(x)\,dx$, $\pi_0(R^n) = 1$. We can now state the following result:

Lemma 5.1 *Consider the optimization problem as stated in the introduction of this section. Let the assumptions (A*1)–(A*4) hold and suppose $(\sigma^0, u^0) \in \Gamma \times \mathscr{U}$ is the optimal policy (whose existence is assumed). Then there exists a weak solution q^0 of the adjoint system AS' corresponding to the policy (σ^0, u^0) such that*

$$\int_Q [(a_{ij}(t, x, \sigma) - a_{ij}(t, x, \sigma^0)).\phi_{x_i}(\sigma, u)(t, x).q^0_{x_j}(t, x)$$
$$+ (a_i(t, x, \sigma, u) - a_i(t, x, \sigma^0, u^0)).\phi_{x_i}(\sigma, u)(t, x).q^0(t, x)]\,dt\,dx$$
$$\leq \int_Q (L(t, x, \sigma, u) - L(t, x, \sigma^0, u^0))q^0(t, x)\,dt\,dx \qquad (5.23)$$

for all $(\sigma, u) \in \Gamma \times \mathscr{U}$.

The proof of this lemma is long and is based on arguments leading to the weak convergence in $L^2(I, W'(R^n, R))$ of the sequence of weak solutions of certain first boundary problems corresponding to the Cauchy problems S' and AS' with their coefficients and data replaced by their integral averages and defined on an increasing sequence of sets $Q_k \equiv I \times \{x : |x| < k\}$ converging to $Q \equiv I \times R^n$. For details the reader is referred to Ahmed and Teo [6, Lemma 6.2]. Using the above lemma, Ahmed and Teo [6, Theorem 6.3] proved the following necessary condition of optimality.

Theorem 5.4 *Consider the optimization problem above. Suppose that the assumptions (A*1)–(A*4) hold and that $a_{ij}(t, x,.)$, $a_i(t, x,.,.)$ $(i, j = 1, 2, \ldots n)$ belong to $C'(\Gamma \times U, R)$, $\Gamma \times U \subset R^{\tilde{q}} \times R^p$ almost everywhere in Q with the gradients bounded for almost all $(t, x) \in Q$ and every $(\sigma, u) \in \Gamma \times U$. Further, it is assumed that the coefficient L is Gateaux differentiable in the weak sense of $L^2(Q, R)$ at each point of $\Gamma \times U$. Then if $(\sigma^0, u^0) \in \Gamma \times \mathcal{U}$ is an optimal policy, it is necessary that there exist a weak solution q^0 of the adjoint system AS' corresponding to the policy (σ^0, u^0) so that for all $(\sigma, u) \in \Gamma \times \mathcal{U}$*

$$\int_Q \langle a_{i,j,\sigma}(\sigma^0)\phi_{xi}(\sigma^0, u^0)q^0_{xj} + a_{i,\sigma}(\sigma^0, u^0)\phi_{xi}(\sigma^0, u^0)q^0 - L_\sigma(\sigma^0, u^0)q^0, \sigma - \sigma^0 \rangle \, dt\, dx$$

$$- \int_Q \langle (f_{i,u}(\sigma^0, u^0)\phi_{xi}(\sigma^0, u^0) + L_u(\sigma^0, u^0))q^0, u - u^0 \rangle \, dt\, dx \leq 0, \quad (5.24)$$

where $h_y(y^0)$ denotes the Gateaux differential of h at y^0 defined by

$$\lim_{\varepsilon \downarrow 0} \frac{h(y^0 + \varepsilon \theta) - h(y^0)}{\varepsilon} = \langle h_y(y^0), \theta \rangle \equiv \sum_i h_{yi}(y^0)\theta_i.$$

The proof of the theorem is obtained from Lemma 5.1 by defining $(\sigma, u) - (\sigma^0, u^0) \equiv (\sigma', u')$, replacing (σ, u) by $(\sigma^0, u^0) + \varepsilon(\sigma', u')$, and dividing the resulting inequality [see the inequality (5.23)] on either side by ε and then letting $\varepsilon \downarrow 0$.

In order to determine the optimal policy it is required to solve the Cauchy problems S' and AS' coupled through the necessary condition of optimality (5.24). It follows from the inequality (5.24) that the necessary condition for optimality for parameters alone is given by

$$\left\langle \int_Q \{a_{i,j,\sigma}(\sigma^0)\phi_{xi}(\sigma^0)q^0_{xj} + a_{i,\sigma}(\sigma^0)\phi_{xi}(\sigma^0)q^0 - L_\sigma(\sigma^0)q^0\} \, dt\, dx, \sigma - \sigma^0 \right\rangle$$

$$\leq 0 \quad \text{for all} \quad \sigma \in \Gamma, \quad (5.25)$$

while for controls alone it is given by the inequality

$$\int_Q \langle [f_{i,u}(u^0)\phi_{xi}(u^0) + L_u(u^0)]q^0, u - u^0 \rangle \, dt\, dx \geq 0 \quad \text{for all} \quad u \in \mathcal{U}. \quad (5.26)$$

It is clear from the necessary condition (5.25) that in case the cost integrand L is independent of σ (or linear in σ), the coefficients a_{ij}, a_i $(i, j = 1, 2, \ldots n)$ are linear in σ, and the parameter space Γ is a polyhedron, the optimal parameter is given by one of the vertices $V(\Gamma)$ of Γ provided the components of the $\tilde{q}$-vector

$$\int_Q [a_{i,j,\sigma}(\sigma^0)\phi_{xi}(\sigma^0)q^0_{xj} + a_{i,\sigma}(\sigma^0)\phi_{xi}(\sigma^0)q^0 - L_\sigma(\sigma^0)q^0] \, dt\, dx$$

are sign definite. In this situation the Cauchy problem S' may be solved for each $\sigma \in V(\Gamma)$ and the optimal parameter determined by comparing the cost integral

$$J(\sigma) = \int_{B \equiv R^n} \phi(\sigma)(0, x) q_0(x) \, dx$$

over $V(\Gamma)$ (Teo et al. [68]).

The necessary condition for optimality for control given by (5.26) is in integral form and can be used in case there is integral constraint rather than amplitude constraint on the control variables. For example, if $\mathcal{U}$ denotes the space of measurable functions on $I \times R^l$ to R^p with the energy constraint $\int_{\tilde{Q}} |u(t, \tilde{x})|^2 \, dt \, d\tilde{x} \leq 1$ and $f(t, x, u) = A(t)x + B(t)u, L(t, x, u) = (K(t, x), u) + K_0(t, x)$, where $K: I \times R^n \to R^p$, $K_0: I \times R^n \to R$, then the optimal control u^0 has the form

$$u^0(t, \tilde{x}) = -\frac{\int_{\tilde{B}} [B'\phi_x^0 + K] q^0 \, d\tilde{\tilde{x}}}{(\int_{\tilde{Q}} |\int_{\tilde{B}} [B'\phi_x^0 + K] q^0 \, d\tilde{\tilde{x}}|^2 \, dt \, d\tilde{x})^{1/2}}, \quad (5.27)$$

where $\tilde{B} = R^l$, $\tilde{\tilde{B}} = R^{n-l}$, $\tilde{Q} = (0, T] \times R^l$. Substituting this expression for u^0 in the Cauchy problems S' and AS' (with σ-deleted), we obtain two coupled nonlinear functional partial differential equations to determine ϕ^0 and q^0 and, in turn, the control u^0.

When the control has amplitude constraints the necessary condition (5.26) cannot be used directly; it is essential to use a pointwise version of this, as given by

$$\left\langle \int_{\tilde{B}} [f_{i,u}(t, \tilde{x}, \tilde{\tilde{x}}, u^0(t, \tilde{x})) \phi_{xi}^0(t, \tilde{x}, \tilde{\tilde{x}}) + L_u(t, \tilde{x}, \tilde{\tilde{x}}, u^0(t, \tilde{x}))] q_c^0 \, d\tilde{\tilde{x}}, v - u^0(t, \tilde{x}) \right\rangle$$

$$\geq 0 \quad \text{for all} \quad v \in U, \quad (5.28)$$

where $q_c^0 = q^0/\hat{q}^0$ and $\hat{q}^0(t, \tilde{x}) = \int_B q^0(t, \tilde{x}, \tilde{\tilde{x}}) \, d\tilde{\tilde{x}}$, $\tilde{\tilde{B}} = R^{n-l}$. For example, if f and L are linear in u as before, and the set U is a unit hypercube in R^p, then the optimal control is given by

$$u^0(t, \tilde{x}) = \text{sgn}\left(\int_{\tilde{B}} (B'\phi_x^0 + K) q_c^0 \, d\tilde{\tilde{x}}\right). \quad (5.29)$$

Substituting this expression in the Cauchy problems S' and AS', again we obtain a coupled system of nonlinear functional partial differential equations which must be solved to determine the optimal control.

Again, if $f = a(t, x) + B(t)u$ (linear in u) and the cost integrand $L = x'Mx + u'Nu$, N symmetric and positive definite, and if the control has energy constraint $\|u\| \leq 1$, then the optimal control u^0 has the form

$$u^0(t, \tilde{x}) = \begin{cases} v(t, \tilde{x}) & \text{for} \quad \|v\| \leq 1 \\ v(t, \tilde{x})/\|v\| & \text{for} \quad \|v\| > 1, \end{cases} \quad (5.30)$$

where $v(t, \tilde{x}) = -\frac{1}{4} \int_{\tilde{B}} (N^{-1} B' \phi_x^0) q_c^0 \, d\tilde{x}$; similarly, in the case of amplitude constraints with the components $|u_i| \leq 1, i = 1, 2, \ldots p$, and f and L as above, the form of the optimal control is given by

$$u_i^0(t, \tilde{x}) = \begin{cases} v_i(t, \tilde{x}) & \text{for } |v_i| \leq 1 \\ v_i(t, \tilde{x})/|v_i(t, \tilde{x})| & \text{for } |v_i| > 1. \end{cases} \quad (5.31)$$

It is clear from the necessary conditions given in this section that, in case only partial information is available for control, one must solve functional partial differential equations to determine the optimal control law. However, in the case of complete information (Section V.B) only partial differential equations are solved. In the former case no suitable computational algorithm is known.

Remark The expression for the optimal control law given by the equation (5.27) is similar to that for linear deterministic systems [36, p. 126] obtained by use of Gamkrelidze's maximum principle of the integral type [36, Theorem 2.1, p. 100]. In fact the expression for optimal control given in [36, p. 126] follows as a special case from the expression (5.27) when the noise is absent ($g \equiv 0$) and π_0 is taken as the Dirac measure at $x_0 \in R^n$, implying $q^0(t, x^0(t)) = \delta(x - x^0(t))$ and $\phi_x^0(t, x^0(t)) = \psi(t)$.

F. A Class of Degenerate Stochastic Systems

In the previous two sections we have considered stochastic control problems in which the diffusion matrix $a = \frac{1}{2} gg'$ is positive definite and consequently the corresponding partial differential equation of dynamic programming is uniformly parabolic. This is known as the nondegenerate case. In most applications the diffusion matrix a is not positive definite and therefore the corresponding partial differential equation is not uniformly parabolic. A natural example of a degenerate problem is a control system described by an nth-order differential equation with white noise input. It is known that nondegenerate parabolic partial differential equations have smooth solutions in the classical sense. Using the theory of distributions, Fleming defined the concept of weak solutions for degenerate Cauchy problems for parabolic partial differential equations [31]. Later [26, Theorem 5.1, p. 269] he used this concept to prove, via the equation of dynamic programming, the existence of weak solutions for degenerate stochastic control systems. We have seen in Section V.B that when the differential equation of dynamic programming (5.7) is uniformly parabolic, it has a smooth solution and that this solution equals the "value function" defined by (5.5). The question that arises is: Does the weak solution of the equation of dynamic programming, corresponding to degenerate stochastic problems, equal the corresponding value function? This was a conjecture of Fleming [26, p. 269] and was later resolved by Rishel [56, pp. 519–528]. He showed, under assumptions similar to those of Fleming [26, p. 268] that if the partial derivatives of the weak solution are functions in appropriate L_p-spaces and solutions of an associated uncontrolled stochastic differential equation have a density which belongs to an appropriate L_p-space, then the weak solution equals the "value function" of the stochastic control problem. This result is briefly presented below. Let n_1, n_2 with $n = n_1 + n_2$ be positive integers and M_{n_2} the space of nonsingular $n_2 \times n_2$ matrices. The system considered by Rishel [56] is governed by the following class of stochastic differential equation in R^n:

$$S \quad \begin{cases} dx_1(t) = f^1(t, x_1(t), x_2(t)) \, dt, & t \in I = [0, T], \\ dx_2(t) = f^2(t, x_1(t), x_2(t), u(t)) \, dt + g(t, x_2(t)) \, dw, & t \in [0, T], \end{cases}$$

where $x_1(t) \in R^{n_1}, x_2(t) \in R^{n_2}, x(t) = (x_1(t), x_2(t)) \in R^n, u(t) \in R^k$ is the control, w is a n_2-vector standard Wiener process on I based on the probability space (Ω, β, P), and

$$f^1: I \times R^{n_1} \times R^{n_2} \to R^{n_1},$$
$$f^2: I \times R^{n_1} \times R^{n_2} \times k^* \to R^{n_2},$$
$$g: I \times R^{n_2} \to M_{n_2}.$$

The assumptions used for the coefficients f^1, f^2, g are

(R1) f^1 is bounded and continuous and satisfies

$$|f^1(t, \xi, \eta) - f^1(t, \xi', \eta')| \leq k_1(|\xi - \xi'| + |\eta - \eta'|)$$

for $(\xi, \eta), (\xi', \eta') \in R^n$ and $t \in I$.

(R2) f^2 is continuous in the control variable and bounded Borel measurable in the rest of the variables.

(R3) g is a bounded continuous M_{n_2}-valued function and g^{-1} is bounded and continuous with values in M_{n_2}.

Admissible Controls Let U be a compact subset of R^k, F denote the space of all functions on I to R^n, and $\mathscr{A}$ denote the Borel field on F generated by the cylinder sets. A function $u: I \times F \to U$ will be called an admissible control if it is nonanticipative and Borel measurable. The class of admissible controls will be denoted by $\mathscr{U}$ as before.

The Optimal Control Problem Let $B \subset R^n$ be a bounded set with compact closure and the boundary $\partial B \in C^{2+\alpha}$, $0 < \alpha < 1$, and let $(t_0, x_0) \equiv (t_0, \xi, \eta) \in I \times B$ be an initial state from which the system S evolves under the action of an admissible control u generating the trajectory $x = (x_1, x_2)$

defined by

$$x_1(t) = \xi + \int_{t_0}^t f^1(s, x_1(s), x_2(s)) \, ds$$
$$x_2(t) = \eta + \int_{t_0}^t f^2(s, x_1(s), x_2(s), u(s)) \, ds + \int_{t_0}^t g(s, x_2(s)) \, dw(s) \quad (5.32)$$

for $t \in [t_0, T]$ and

$$u(t) = v(t, x_1(.), x_2(.)). \quad (5.33)$$

Let $\tau \equiv \inf\{t \in [t_0, T] : x(t) = (x_1(t), x_2(t)) \in \partial B\}$ be the stopping time for the process x and $\tau_1 = \tau \wedge T$. The optimal control problem is to find a control law $u \in \mathcal{U}$ such that

$$\phi^u(t_0, \xi, \eta) = E \int_{t_0}^{\tau_1} L(s, x_1(s), x_2(s), v(s, x_1(.), x_2(.))) \, ds \quad (5.34)$$

is a minimum.

Solution of the Control Problem Since f^2 is assumed to be merely Borel measurable the question of existence of a solution to (5.32) cannot be settled by classical means. At this point Rishel uses Girsanov's procedure (Section III) and gives an interesting proof for the existence of a solution to (5.32) [56, Theorem 1, p. 521]. As an outline of his proof let us denote by $\tilde{P}$ the measure given by

$$d\tilde{P} = \exp[\zeta_{t_0}^T(f^2)] \, dP, \quad (5.35)$$

where

$$\zeta_{t_0}^t(f^2) = \int_{t_0}^t f^{2\prime}(s, x_1(s), x_2(s), u(s))[g'(s, x^2(s))]^{-1} \, dw(s)$$
$$- \tfrac{1}{2} \int_{t_0}^t |[g(s, x_2(s))]^{-1} f^2(s, x_1(s), x_2(s), u(s))|^2 \, ds; \quad (5.36)$$

$u(s) = v(s, x_1, x_2)$ and (x_1, x_2) is the solution of the uncontrolled system

$$x_1(t) = \xi + \int_{t_0}^t f^1(s, x_1(s), x_2(s)) \, ds,$$
$$x_2(t) = \eta + \int_{t_0}^t g(s, x_2(s)) \, dw(s), \quad t \in [t_0, T]. \quad (5.37)$$

Then one shows that the process $\{\tilde{w}(t) : t \in I\} \equiv \tilde{w}$ defined by

$$\tilde{w}(t) = w(t) - \int_{t_0}^t g^{-1}(s, x_2(s)) f^2(s, x_1(s), x_2(s), u(s)) \, ds \quad (5.38)$$

with respect to the probability space $(\Omega, \beta, \tilde{P})$ is a Brownian motion process and the solution (x_1, x_2) of (5.37), when considered with respect to the probability space $(\Omega, \beta, \tilde{P})$, is the solution of the original system (5.32) with $\tilde{w}$ replacing w.

Using this fundamental result, essentially due to Grisanov [35, Theorem 1, p. 287], Rishel showed the equivalence of the value function of the stochastic control problem and the solution of certain quasi-linear parabolic differential equations.

For $\gamma > 1$ let F_γ denote the class of all continuous real-valued functions defined on $I \times B$ such that their first partials with respect to (t, ξ) and second partials with respect to η, considered in the sense of distribution, belong to $L_\gamma(I \times B, R)$. Define a norm on F_γ by

$$|\phi| \equiv \sup_{(t,\xi,\eta) \in I \times B} |\phi(t, \xi, \eta)| + \sum_{z=t,\xi_i,\eta_i} \left(\int_{I \times B} |\phi_z|^\gamma \, dt \, d\xi \, d\eta \right)^{1/\gamma}$$
$$+ \sum_{i,j=1}^{n_2} \left(\int_{I \times B} |\phi_{\eta_i,\eta_j}|^\gamma \, dt \, d\xi \, d\eta \right)^{1/\gamma}. \tag{5.39}$$

For $u \in U$ define the differential operator in $I \times B$

$$A(t, \xi, \eta, u)[\phi] = \Lambda(\phi) + \sum_{i=1}^{n_2} f_i^2(t, \xi, \eta, u) \phi_{\eta_i},$$

where

$$\Lambda(\phi) = \phi_t + \sum_{i,j=1}^{n_2} a_{ij}(t, \eta) \phi_{\eta_i,\eta_j} + \sum_{i=1}^{n_1} f_i^1(t, \xi, \eta) \phi_{\xi_i},$$

and consider the quasi-linear parabolic differential equation with first boundary condition

$$\Lambda(\phi) + \hat{H}(t, \xi, \eta, \phi_\eta) = 0, \quad (t, \xi, \eta) \in I \times B,$$
$$\phi(t, \xi, \eta) = 0 \quad \text{for} \quad (t, \xi, \eta) \in \Sigma \equiv (I \times \partial B) U \{T\} \times B, \tag{5.40}$$

where

$$\hat{H}(t, \xi, \eta, z) = \min_{u \in U} \left\{ \sum_{i=1}^{n_2} f_i^2(t, \xi, \eta, u) z_i + L(t, \xi, \eta, u) \right\}. \tag{5.41}$$

A function ϕ is said to be a weak solution of the first boundary problem (5.40) if $\phi \in F_\gamma$ and it satisfies (5.40) a.e. with respect to $(1 + n)$-dimensional Lebesgue measure on $I \times B$. Under the assumption that the uncontrolled system (5.37) has a solution (x_1, x_2) with random vector $(x_1(t), x_2(t))$ having a density $\pi \equiv (\pi(t, \xi, \eta)) \in L_p(Q_\delta, R)$ for some $p > 1$, where $Q_\delta = [t_0 + \delta, T] \times B$, Rishel proved [57, Lemma 1, p. 522] that the random vector $(x_1(t), x_2(t))$, corresponding to a solution of (5.32), has a density θ that belongs to $L_{p'}(Q_\delta, R)$ for some $1 < p' < p$. Using this result, Ito's lemma, and the fact that the equation of dynamic programming (5.40) has a weak solution $\phi \in F_\gamma$ for

$\gamma = p'(p'-1)^{-1}$ [31, Theorem 5, p. 1001], it can be shown that

$$E\{\phi(\tau_1, x_1(\tau_1), x_2(\tau_1)) - \phi(\tau_\delta, x_1(\tau_\delta), x_2(\tau_\delta))\}$$
$$= \int_{Q_\delta} E\{\chi(s)A(s, x_1(s), x_2(s), v(s, x_1, x_2))[\phi] | x_1(s)$$
$$= \xi, x_2(s) = \eta\}\theta(s, \xi, \eta)\,ds\,d\xi\,d\eta, \qquad (5.42)$$

where $\tau_\delta = \tau_1 \wedge (t_0 + \delta)$, $\chi(s) = 1_{\{\tau_1 \geq s\}}$, and (x_1, x_2) is the solution of (5.32) corresponding to the control u with $u(t) = v(t, x_1, x_2)$ and θ is the corresponding density function. If ϕ is a weak solution of (5.40), it follows from an implicit function theorem due to Beneš [13, Lemma 6, p. 460] that there exists a measurable function $v^* : I \times B \to U$ such that

$$\sum_{i=1}^{n_2} f_i^2(t, \xi, \eta, v^*(t, \xi, \eta))\phi_{\eta_i} + L(t, \xi, \eta, v^*(t, \xi, \eta))$$
$$= \hat{H}(t, \xi, \eta, \phi_\eta) \qquad \text{a.e. on } I \times B. \qquad (5.43)$$

Clearly, this function induces an $\mathscr{A}$-measurable function on $I \times F$ by the formula

$$v^*(t, x_1, x_2) = v^*(t, x_1(t), x_2(t)). \qquad (5.44)$$

Using these results Rishel proved that the value function

$$\psi(t, \xi, \eta) \equiv \inf_{u \in \mathscr{U}} E\left\{ \int_t^{\tau_1} L(s, x_1(s), x_2(s), v(s, x_1, x_2))\,ds \,\Big|\, x_1(t) = \xi, x_2(t) = \eta \right\}$$

equals the solution of the first boundary problem (5.40). More precisely, the following result holds.

Theorem 5.5 (Rishel [56, Theorem 3, p. 526]) *Let the solution of the uncontrolled system (5.37) have a density $\pi \in L_p(Q_\delta, R)$ for some $p > 1$ and every $\delta > 0$. Let $1 < p' < p$ and let ϕ be a weak solution of (5.40) which belongs to F_γ for $\gamma = p'(p'-1)^{-1}$. Let v^* be the control as defined by (5.43) and (5.44).*

Then, if (x_1, x_2) is a solution of (5.32) corresponding to the control $u^*(t) = v^*(t, x_1, x_2)$,

$$\phi(t_0, \xi, \eta) = E\left\{ \int_{t_0}^{\tau_1} L(s, x_1(s), x_2(s), u^*(s))\,ds \right\}$$

and for any other admissible $\mathscr{A}$-measurable control v with corresponding responses (y_1, y_2) of the system (5.32)

$$\phi(t_0, \xi, \eta) \leq E\left\{ \int_{t_0}^{\tau_1} L(s, y_1(s), y_2(s), v(s, y_1, y_2))\,ds \right\}.$$

Thus, $(t_0, \xi, \eta) \in I \times B$ being arbitrary, it follows from the above theorem that

$$\phi(t, \xi, \eta) = \psi(t, \xi, \eta) \quad \text{on} \quad I \times B.$$

In fact Rishel's result provides both a necessary and sufficient condition for optimality and consequently the optimal control law is determined by solving the system of equations (5.40) and (5.41). The assumptions under which the above results have been proved are (i) the controller is based on complete information about the current state $(x_1(t), x_2(t))$, (ii) the response of the uncontrolled system (5.37) has a density $\pi \in L_p(Q, R)$ for some $p > 1$, (iii) f^1 is independent of control u, (iv) g is independent of x_1, and (v) $g(t, x_2)$, $(t, x_2) \in I \times R^{n_2}$, is nonsingular with g^{-1} continuous for $(t, x_2) \in I \times R^{n_2}$. Among these (i) and (iii) are most crucial and have limited application.

G. A Class of Random Differential Systems with Jump Markov Disturbances

In the preceding sections we have mainly considered stochastic systems described by Ito differential equations. In applications often the physical system is modeled as an ordinary differential equation with randomly varying parameters. The parameters may take values from a finite set S, constituting the state space, and the transition from one state to another takes place by jumps. If this transition is assumed to be Markovian, then the parameter process is completely defined by its (one step) transition probability matrix (4.12) as given in Section IV.C. There we considered a linear system in which both the plant and control matrices A and B were assumed to be a finite state Markov process. The linearity restriction is not essential. Consider the interval $I \equiv [0, \infty)$ and let $\{\gamma(t) : t \in I\}$ denote a finite state Markov jump process as described above with state space S and suppose the controlled system is governed by the differential equation

$$d\xi/dt = f(t, \xi(t), \gamma(t), u(t, \xi(t), \gamma(t))), \quad t \in I,$$
$$\xi(0) = x \in R^n, \tag{5.45}$$

where

$$f : I \times R^n \times S \times U \to R^n,$$
$$u : I \times R^n \times S \to U \subset R^m.$$

The function $(t, x, s, u) \to f(t, x, s, u)$ is assumed to be C' in x and u on $R^n \times U$ for each $t \in I$ and $s \in S$ and measurable in t on I for each $x \in R^n$, $u \in U$ and $s \in S$. The function u defined on $I \times R^n \times S$ is a control law with values in U. The terminal conditions may be specified by a manifold $M = \{(t, x) \in$

OPTIMAL CONTROL OF STOCHASTIC SYSTEMS 59

$I \times R^n : \phi_i(t, x) = 0, i = 1, 2, \ldots k\}$ with ϕ_i continuous on $I \times R^n$. Let $\mathcal{U}_M$ denote the class of admissible controls (control laws) so that for each $u \in \mathcal{U}_M$ the system (5.45) has a solution ξ_u and there is a random time $\tau \in I$ such that $\xi_u(\tau) \in M$ with probability one. It is known from deterministic control theory that the optimal control is often discontinuous. Thus it is essential to choose the class $\mathcal{U}_M$ to be large enough to admit such discontinuous controls. In fact it is convenient to assume that for each $(t, x) \in G \equiv I \times R^n$, $\gamma^i \in S$, and control $u \in \mathcal{U}_M$, the deterministic differential equation

$$d\xi^i/d\theta = f(\theta, \xi^i, \gamma^i, u(\theta, \xi^i, \gamma^i)), \quad \theta \in [t, \infty),$$

with initial condition $\xi^i(t) = x$ has a unique solution $\xi^i(\theta)$ on an interval $[t, t^i]$ with $\xi^i(\theta) \notin M$ for $t \leq \theta < t^i$ and, at time $\theta = t^i$, $\xi^i(t^i) \in M$. Under this assumption a sample function interpretation of the solution of the differential equation easily follows. Let (Ω, B, P) be the supporting probability space for the process γ and let $\gamma(.,\omega)$, $\omega \in \Omega$, be any sample path. Since γ is a jump Markov process, for any interval $[t, \infty)$, there exists a divergent sequence of numbers $0 \leq t = s_1 < s_2 < \cdots < s_{m-1} < \cdots < s_m \cdots$ depending on the ω at which the trajectory $\gamma(.,\omega)$ makes jumps and is found in states $\gamma^1, \gamma^2, \ldots, \gamma^m, \ldots \in S$. The corresponding solution $\xi(.,\omega)$ of the differential equation is constructed by piecing together the solutions of the sequence of deterministic differential equations

$$d\xi^1/d\theta = f(\theta, \xi^1(\theta), \gamma^1, u(\theta, \xi^1(\theta), \gamma^1)), \quad \theta \in (t, s_2],$$
$$\xi^1(t) = x, \tag{5.46}$$

and

$$d\xi^i/d\theta = f(\theta, \xi^i(\theta), \gamma^i, u(\theta, \xi^i(\theta), \gamma^i)), \quad \theta \in (s_i, s_{i+1}],$$
$$\xi^i(s_i) = \xi^{i-1}(s_i), \quad i = 2, 3, \ldots \quad \text{(initial condition)}, \tag{5.47}$$

until the process terminates in the terminal set M.

Control Problem Let $f_0 : I \times R^n \times S \times U \to R_+ = [0, \infty)$ and suppose the performance index is given by

$$J(u) = E\left\{\int_0^\tau f_0(s, \xi(s), \gamma(s), u(s, \xi(s), \gamma(s))) \, ds\right\}, \tag{5.48}$$

where ξ is the response of the system (5.45) corresponding to the control law u and τ is the first time $\xi(\tau) \in M$. The problem is to find a control in $\mathcal{U}_M$ that imparts a minimum to the functional J. The function $(t, x, \gamma, u) \to f_0(t, x, \gamma, u)$ defined on $I \times R^n \times S \times U$ is also assumed to be C' in x, u for each $t \in I$ and $\gamma \in S$ and measurable in t for each x, γ, u in their domain of definition. Let us index the set S by $\{i = 1, 2, \ldots k\}$ and denote by $V(t, x, i)$

the "value function" of the optimization problem

$$V(t, x, i) = \inf_{u \in \mathcal{U}_M} E\left\{\int_t^\tau f_0(s, \xi(s), \gamma(s), u(s, \xi(s), \gamma(s))) \, ds \,\Big|\, \xi(t) = x, \gamma(t) = \gamma^i\right\}. \tag{5.49}$$

A control u^* is said to be optimal if

$$E\left\{\int_t^\tau f_0(s, \xi^*(s), \gamma(s), u^*(s, \xi^*(s), \gamma(s))) \, ds \,\Big|\, \xi^*(t) = x, \gamma(t) = \gamma^i\right\} = V(t, x, i), \tag{5.50}$$

where ξ^* is the response of the system corresponding to the control u^*. By the usual arguments of dynamic programming one can easily prove the following necessary condition of optimality.

Theorem 5.6 Let $u^* = u^*(t, x, i)$ be an optimal control. Then at each point $(t, x) \in G$ at which the first partial of $V = V(t, x, i)$ exists and $u^*(t, x, i)$ is continuous, the equation

$$\inf_{u \in U} \left\{ V_t(t, x, i) + V_x(t, x, i) \cdot f(t, x, i, u) + f_0(t, x, i, u) + \sum_{j=1}^k q_{ij} V(t, x, j) \right\}$$

$$= V_t(t, x, i) + V_x(t, x, i) \cdot f(t, x, i, u^*(t, x, i)) + f_0(t, x, i, u^*(t, x, i))$$

$$+ \sum_{j=1}^k q_{ij} V(t, x, j) = 0 \tag{5.51}$$

is satisfied, $i = 1, 2, \ldots k$.

The equation of dynamic programming (5.51) is said to have a generalized or weak solution if the indicated first partials are defined in the sense of distributions and the equality holds almost everywhere in $G = I \times R^n$ with respect to $(n + 1)$-dimensional Lebesgue measure. Using this concept of weak solution, Rishel established the following sufficient condition of optimality.

Theorem 5.7 (Rishel [58, Theorem 2]) *If $W \equiv W(t, x, i)$ is a weak solution of the system of partial differential equations* (5.51) *of dynamic programming with $W(t, x, i) = 0$ for $(t, x) \in M$ and $u \equiv u(t, x, i) \in \mathcal{U}_M$ is such that*

$$E\left\{\int_t^\tau f_0(\theta, \xi(\theta), \gamma(\theta), u(\theta, \xi(\theta), \gamma(\theta))) \, d\theta \,\Big|\, \xi(t) = x, \gamma(t) = \gamma^i\right\} = W(t, x, i), \tag{5.52}$$

then u is an optimal control with ξ the corresponding response.

The proof of this theorem is based on a fundamental inequality which states that if W is a generalized solution of the system of partial differential equations of dynamic programming, then for any arbitrary control $u \in \mathcal{U}_M$

$$W(t, x, i) \leq E\left\{\int_t^\tau f_0(s, \xi(s), \gamma(s), u(s, \xi(s), \gamma(s))) \, ds \, \middle| \, \xi(t) = x, \gamma(t) = \gamma^i\right\} \quad (5.53)$$

for $(t, x, \gamma^i) \in I \times R^n \times S$. Thus a control for which the equality holds is necessarily optimal.

As mentioned in Section IV.C the stochastic maximum principle leads to a system of matrix Riccati differential equations that solves the linear regulator problem with quadratic cost. The same result can be obtained using the dynamic programming approach provided $U = R^m$. Assuming that $\gamma(t) = [A(t), B(t)]$ and, for $\gamma^i \in S, \gamma^i = [A_i, B_i]$, the partial differential equation of dynamic programming is given by

$$\inf\left\{V_t(t, x, i) + (V_x(t, x, i), (A_i x + B_i u)) + (x, Rx) + (u, Su)\right.$$
$$\left. + \sum_j q_{ij}(t)V(t, x, i)\right\} = 0. \quad (5.54)$$

From this expression the control law is found to be

$$u = -\tfrac{1}{2}S^{-1}B_i' V_x(t, x, i). \quad (5.55)$$

Substituting a trial solution

$$V(t, x, i) = (x, \beta(t, i)x) \quad (5.56)$$

in (5.54) and (5.55), one obtains the desired matrix Riccati differential equation for β,

$$\beta_t(t, i) = -\beta A_i - A_i'\beta + \beta(B_i S^{-1} B_i')\beta - R - \sum_j q_{ij}(t)\beta(t, j),$$
$$\beta(T, i) = 0, i = 1, 2, \ldots, \quad t \in [0, T]. \quad (5.57)$$

The optimal control is then given by

$$u = -S^{-1}B_i'\beta(t, i)x \quad \text{if} \quad \xi(t) = x,$$
$$\gamma(t) = [A_i, B_i] \quad \text{at time} \quad t.$$

It is clear that use of the minimum principle instead of the maximum principle used to derive (4.23) would eliminate the apparent discrepancies in signs of the two equations (4.23) and (5.27).

Recently, Rishel has further generalized this result, giving several interesting necessary and sufficient conditions for optimality [57, Theorems 7, 9, 11, and 12]. These results are proved under weaker assumptions on the function f. In particular, f can be discontinuous in both time and the space variable. This permits inclusion of discontinuous controls in the class $\mathcal{U}_M$.

The risk function to be minimized is taken to be

$$J(u) = E\{\psi(\tau_u, \xi(\tau_u)) | \xi(t) = x, \gamma(t) = \gamma^i\},$$

where τ_u is the first time the solution of the system

$$d\xi/ds = f(s, \xi(s), \gamma(s), u(s, \xi(s), \gamma(s)))$$
$$\xi(t) = x, \quad s \in [t, \infty),$$

corresponding to the control $u \in \mathcal{U}_M$, reaches a closed terminal manifold $M \subset G$, G an open set in $[0, \infty) \times R^n$. Under several technical conditions it has been shown that for each control law $u \in \mathcal{U}_M$ the function

$$V(t, x, i) \equiv E\{\psi(\tau_u, \xi(\tau_u)) | \xi(t) = x, \gamma(t) = \gamma_i\} \quad (i = 1, 2, \ldots k)$$

and its partials V_x and V_t satisfy certain integral equations [57, Theorems 2 and 3, Corollary 2] of Volterra type, $V(t, x, i)$ is continuous in x and t in the region where $f(t, x, i, u(t, x, i))$ is, and that V_t and V_x exist and are locally bounded.

Using these results and the maximum principle [57, Theorem 6] for partial differential equations of the type

$$h^i_t(t, x) + h^i_x(t, x) \cdot f^i(t, x) + \sum_{j=1} q_{ij} h^j(t, x) = 0,$$

$$(t, x) \in G \backslash S^n, \quad h^i(t, x) = \psi(t, x), \quad (t, x) \in M \subset G,$$

Rishel has established [57, Theorem 7] that the necessary and sufficient conditions for optimality of a control law $u = u(t, x, i) \in \mathcal{U}_M$ is that the corresponding cost function $V = V(t, x, i)$ satisfies the system of partial differential equations of dynamic programming (5.51) on $G \backslash S_i^n$ with boundary condition $V(t, x, i) = \psi(t, x)$ for $(t, x) \in M$. The set S_i^n is the set of discontinuities of the function $f(t, x, i, u(t, x, i))$ in G. It is interesting to note that the proof of this result has been substantially simplified by use of the integral equations and the maximum principle mentioned above. Further, using the integral equation satisfied by the partial V_x of the cost function V, Rishel was able to transform the necessary and sufficient conditions of dynamic programming [57, Theorem 7] into the necessary and sufficient conditions [57, Theorems 9, 11, and 12] which are analogous to those of Pontryagin [54, Theorem 22, p. 267] for deterministic control problems with state space constraints. These later necessary and sufficient conditions contain both transversality and jump conditions as found in deterministic control problems. For a detailed proof of these interesting results and their usefulness in the actual computation of optimal controls the reader is referred to the original work of Rishel [57].

Recently, Ahmed and Wong [10] have presented a necessary condition of optimality for a more general problem:

$$d\xi(t) = f(t, \xi(t), \gamma(t), u(t, \xi(t), \gamma(t))) \, dt + g(t, \xi(t), \gamma(t))) \, dw(t),$$

W $\qquad \xi(0) = \xi_0$ with initial probability density q_0,

$$J(u) = E \int_0^\tau L(t, \xi(t), \gamma(t), u(t, \xi(t), \gamma(t))) \, dt = \min,$$

where γ is a finite state Markov jump process with states indexed by a finite set of integers $\{1, 2, \ldots N\}$, τ is the first exit time of ξ from the cylinder $Q_T = (0, T) \times \Sigma$, Σ a bounded open subset of R^n. The control class $\mathcal{U}$ consists of measurable functions on Q_T with values in a compact convex subset of a finite-dimensional space. For convenience of notation for each $k \in \{1, 2, \ldots N\}$

$$b_i^u(t, x, k) = f_i(t, x, k, u(t, x, k)), L^u(t, x, k) = L(t, x, k, u(t, x, k))$$

and

$$a_{ij}(t, x, k) = \tfrac{1}{2}(g(t, x, k) \cdot g'(t, x, k))_{ij}, \qquad i, j = 1, 2, \ldots, n, \quad k = 1, 2, \ldots, N.$$

Let $A(u)$ denote the $(N \times N)$ matrix of partial differential operators with elements

$$(A(u))_{kl} \equiv \begin{cases} q_{k,l}(t) & \text{for } k \neq l \\ \sum_{i,j} a_{ij}(t, x, l) \dfrac{\partial^2}{\partial x_i \partial x_j} + \sum_i b_i^u(t, x, l) \dfrac{\partial}{\partial x_i} + q_{ll} & \text{for } k = l; \end{cases}$$

$$L(u) \equiv (L^u(t, x, 1), \ldots, L^u(t, x, N))',$$
$$\Phi(u) \equiv (\Phi(u)(t, x, 1), \ldots, \Phi(u)(t, x, N))',$$
$$\Psi(u) \equiv (\Psi(u)(t, x, 1), \ldots, \Psi(u)(t, x, N))',$$

and

$$\bar{B}_i(u) \equiv \operatorname{diag}\left\{b_i^u(t, x, l) - \sum_j (\partial a_{ji}/\partial x_j)(t, x, l) : l = 1, 2, \ldots N\right\}.$$

Let $P^0 = (p_1^0, \ldots, p_n^0)$ be the initial distribution of $\gamma(0)$. The necessary conditions are given in the following result.

Theorem 5.8 *For every control $u \in \mathcal{U}$, let a_{ij} be bounded continuous and positive definite, $\bar{B}_i(u)$ belong to $L_{r,q}(Q_T)$, $L(u)$ belong to $L_{r_1,q_1}(Q_T)$, and let $\bar{B}_i(u)$, $L(u)$ be dominated by some functions in $L_{r,q}(Q_T)$ and $L_{r_1,q_1}(Q_T)$, respectively. Further, suppose the strong Gateaux differentials $\bar{B}_{iu}$, L_u belong, respectively, to the same function spaces as those of $\bar{B}_i$ and L. Let u^0 be the optimal control.*

Then the necessary conditions of optimality consist of the following systems of parabolic equations and the integral inequality:

(i)
$$\begin{cases} \Phi_t + A(u^0)\Phi + L(u^0) = 0 & \text{for } (t,x) \in [0,T) \times \Sigma, \\ \Phi(t,x) = 0 & \text{for } (t,x) \in [0,T] \times \partial\Sigma, \\ \Phi(T,x) = 0 & \text{for } x \in \Sigma; \end{cases}$$

(ii)
$$\begin{cases} -\Psi_t + A^*(u^0)\Psi = 0 & \text{for } (t,x) \in (0,T] \times \Sigma, \\ \Psi(t,x) = 0 & \text{for } (t,x) \in [0,T] \times \partial\Sigma, \\ \Psi(0,x) = q_0(x)P^0 & \text{for } x \in \Sigma; \end{cases}$$

(iii)
$$\int_{Q_T} \left\langle \sum_i \bar{B}_{iu}(u - u^0) \cdot \Phi_{x_i}(u^0) - L_u(u - u^0), \Psi(u^0) \right\rangle dx\,dt \leq 0$$

for all $u \in \mathcal{U}$.

The proof of this result [10] is based on the existence of weak solutions of parabolic systems in the divergence form and their continuity properties with respect to controls in some suitable topology. Variational technique is employed to obtain the inequality (iii). One principal advantage of the above necessary conditions of optimality is that it allows computation of the optimal control by successive application of a simple gradient technique.

H. Some Open Problems

For practical applications the most important results desired are the necessary conditions of optimality that can be used for synthesis of optimal feedback control laws. We briefly mention here some interesting open problems in this area.

(i) Necessary conditions of optimality admitting controls measurable in both time and space variable and appearing in both the local drift and the diffusion coefficients will be of considerable practical interest. No such necessary conditions appear to have been reported in the literature.

(ii) It would be useful to develop necessary conditions of optimality for systems governed by nonlinear Ito differential equations equipped with a data storage device that can store incomplete state information over specified lengths of time (in the past) and make it available for control computation.

(iii) Necessary conditions of optimality for systems governed by generalized Ito stochastic differential equations [60, p. 45] that include random Poisson jump measure are not known.

(iv) It would be interesting to develop a computational technique for solving a coupled system of integropartial differential equations like (5.21) and (5.22) that arise from the necessary conditions of optimality for partially observed diffusions.

(v) For successful application of known results in the synthesis of optimal control law it is necessary to develop an efficient algorithm for solving the equations of dynamic programming in spaces of dimension larger than two.

ACKNOWLEDGMENTS

This article is based on a graduate course given by the author at the Department of Electrical Engineering of the University of Ottawa under the title "Stochastic Optimal Controls" during the years 1971–1975. The author would like to thank his graduate students, especially Dr. K. L. Teo, now with the Department of Mathematics of the University of New South Wales, and Mr. H. W. Wong, who made numerous suggestions for improvement of the presentation.

REFERENCES

1. Ahmed, N. U., A class of stochastic non-linear integral equations on L_p space and its application to optimal control, *J. Information and Control* **14** (1969), 512–523.
2. Ahmed, N.U., Certain topological properties of stochastic processes generated by a family of stochastic Ito-differential systems, *Ricerche Automatica* **4** (1973), 1–13.
3. Ahmed, N. U., Optimal control of stochastic dynamical systems, *J. Information and Control* **22** (1973), 13–30.
4. Ahmed, N. U., Strong and weak synthesis of non-linear systems with constraints on the system space G , *J. Information and Control* **23**, No. 1 (1973), 71–85.
5. Ahmed, N. U., and Teo, K. L., An existence theorem on optimal control of partially observable diffusions, *SIAM J. Control* **12**, No. 3 (1974), 351–355.
6. Ahmed, N. U., and Teo, K. L., Optimal control of stochastic Ito-differential systems with fixed terminal time, *Advances in Appl. Probability* **7**, March (1974), 1–25.
7. Ahmed, N. U., and Teo, K. L., Optimal control of stochastic McShane differential systems, *J. Appl. Probability* **11**, No. 2 (1974), 302–309.
8. Ahmed, N. U., and Teo, K. L., Stochastic Bang-bang control, *IEEE Trans Automatic Control* **AC-19**, No. 1 (1974), 73–75.
9. Ahmed, N. U., and Georganas, N. D., On optimal parameter selection, *IEEE Trans. Automatic Control* **AC-18**, No. 3 (1973), 313–314.
10. Ahmed, N. U., and Wong, H. W., A minimum principle for systems governed by Ito differential equations with Markov jump parameters, in "Differential Games and Control Theory II" (E. O. Roxin, P. T. Liu, and R. L. Sternberg, eds.), Lecture Notes in Pure and Applied Mathematics, Vol. 30. Dekker, New York, 1977, pp. 265–294.
11. Becker, H., and Mandrekar, V., On the Existence of Optimal Random Controls, MRC Tech. Summary Rep. No. 895, 1968.
12. Benés, V. E., Existence of optimal stochastic control laws, *SIAM J. Control* **9** (1971), 446–472.
13. Benés, V. E., Existence of optimal strategies based on specified information for a class of stochastic decision problems, *SIAM J. Control* **8** (1970), 179–188.
14. Bensoussan, A., and Voit, M., Optimal control of stochastic linear distributed parameter Systems, *SIAM J. Control* **13**, No. 4 (1975), 905–926.

15. Bharucha-Reid, A. T., Random algebraic equations, *in* "Probabilistic Methods in Applied Mathematics" (A. T. Bharucha-Reid, ed.), Vol. 1. Academic Press, New York, 1970, pp. 1–52.
16. Bharucha-Reid, A. T., On random solutions of integral equations in banach space, *Trans. 2nd Prague Confer. Information Theory, Stochastic Decision Functions, Random Processes* (1960), 27–48.
17. Bharucha-Reid, A. T., "Random Integral Equations." Academic Press, New York, 1972, pp. 218–258.
18. Bismut, J. M., Linear quadratic optimal stochastic control with random coefficients, *SIAM J. Control* **14**, No. 3 (1976), 419–444.
19. Bismut, J. M., On optimal control of linear stochastic equations with linear quadratic criterion, *SIAM J. Control* **15**, No. 1 (1977), 1–4.
20. Bucy, R. S., and Joseph, P. D., "Filtering for Stochastic Processes with Applications to Guidance. Wiley (Interscience), New York, 1968, pp. 65–72.
21. Cesari, L., Existence theorems for weak and usual optimal solutions in Lagrange problems with unilateral constraints, *Trans. Amer. Math. Soc.* **124** (1966), 369–412.
22. Davis, M. H. A., and Varaiya, P., Dynamic programming conditions for partially observable stochastic systems, *SIAM J. Control* **11**, (1973), 226–261.
23. Duncan, T. E., and Varaiya, P., On the solutions of stochastic control systems, *SIAM J. Control* **9** (1971), 354–371.
24. Dunford, N., and Schwartz, J. T., "Linear Operators," Part 1, Wiley (Interscience), New York, 1964, p. 426.
25. Fleming, W. H., "Controlled diffusions under polynomial growth conditions, *in* "Control Theory and Calculus of Variations" (A. V. Balakrishnan, ed). Academic Press, New York, 1969, pp. 209–234.
26. Fleming, W. H., Duality and *a priori* estimates in Markovian optimization problems, *J. Math. Anal. Appl.* **16** (1966), 254–279.
27. Fleming, W. H., Dynamical systems with small stochastic terms, *in* "Techniques of Optimization" (A. V. Balakrishnan, ed.). Academic Press, New York, 1972, pp. 325–334.
28. Fleming, W. H., Optimal control of diffusion processes, *in* "Functional Analysis and Optimization" (E. R. Caianiello, ed.). Academic Press, New York, 1966, pp. 65–84.
29. Fleming, W. H., Optimal continuous parameter stochastic control, *SIAM Rev.* **11**, No. 4 (1969), 470–509.
30. Fleming, W. H., Optimal control of partially observable diffusions, *SIAM J. Control* **6** (1968), 194–213.
31. Fleming, W. H., The Cauchy problem for degenerate parabolic equations, *J. Math. Mech.* **13** (1964), 987–1008.
32. Fleming, W. H., and Nisio, M., On the existence of optimal stochastic controls, *J. Math. Mech.* **15** (1966), 777–794.
33. Florentine, J. J., Optimal control of continuous time Markov stochastic systems, *J. Electron. Control* **10** (1961), 473–488.
34. Gihman, I. I., and Skorokhod, A. V., "Stochastic Differential Equations." Springer-Verlag, Berlin and New York, 1972.
35. Girsanov, I. V., On transforming a certain class of stochastic processes by absolutely continuous substitution of measures, *Theor. Probability Appl.* **5** (1960), 285–301.
36. Gramkrelidze, R. V., On some extremal problems in the theory of differential equations with application to optimal control, *SIAM J. Control* **3**, No. 1 (1965), 107–128.
37. Guinn, T., Solutions of generalized optimization problems, *J. Comput. System Sci.* (1967), 227–334.
38. Himmelberg, C. J., Jacobs, M. Q., and Van Vleck, F. S., Measurable multifunctions,

selectors and Filippov's implicit functions lemma, *J. Math. Anal. Appl.* **25** (1969), 276–284.
39. Holtzman, J. M., On the maximum principle for non-linear discrete time systems, *IEEE Trans. Automatic Control* **AC-4** (1966), 528–547.
40. Joseph, P. D., and Tou, J. T., On linear control theory, *AIEE Trans. Application Industry* **80** (1961), 193–196.
41. Kushner, H. J., Near optimal control in the presence of small stochastic perturbations, *J. Basic Energy.* **87** (1965), 103–108.
42. Kushner, H. J., Necessary conditions for continuous parameter stochastic optimization problems, *SIAM J. Control* **10** No. 3 (1972), 550–565.
43. Kushner, H. J., Necessary conditions for discrete parameter stochastic optimization problems, *Proc. 6th Berkeley Symp. Math., Statist. Probability.*
44. Kushner, H. J., On the existence of optimal stochastic controls, *SIAM J. Control, Ser. A*, **3** (1966), 463–474.
45. Kushner, H. J., On stochastic extremum problems, calculus, *J. Math. Anal. Appl.* **10** (1965), 354–367.
46. Kushner, H. J., On the stochastic maximum principle; fixed time of control, *J. Math. Anal. Appl.* **11** (1965), 78–92.
47. Kushner, H. J., On the stochastic maximum principle with "average" constraints, *J. Math. Anal. Appl.* **12** (1965), 13–26.
48. Kushner, H. J., and Kleinman, A. I., Mathematical programming and the control of Markov chains, *Internat. J. Control* **13** (1971), 801–820.
49. Ladyzhenskaya, O. A., Solonikov, V. A., and Uralcéva, N. N., Linear and quasilinear equations of parabolic type, *Engrg. Trans., Amer. Math. Soc.* (1968), 492–496.
50. McShane, E. J., and Warfield, R. B., Jr., On Filippov's implicit functions lemma, *Proc. Amer. Math. Soc.* **18** (1967), 41–47.
51. McShane, E. J., Relaxed controls and variational problems, *SIAM J. Control* **5** (1967), 438–485.
52. Neustadt, L. W., An abstract variational theory with applications to a broad class of optimization problems, *SIAM J. Control* **4** (1966), 505–527.
53. Oğuztöreli, M. N., "Time-Lag Control Systems." Academic Press, New York, 1966, pp. 181–191.
54. Pontryagin, L. S., Boltyanskii, V. G., Gamkrelidze, R. V., and Mishchenko, E. F., "The Mathematical Theory of Optimal Processes." Wiley (Interscience), New York, 1965.
55. Potter, J. E., A guidance–Navigation Separation Theorem, Rep. Re-11, Experimental Astronomy Laboratory, M.I.T., Cambridge, Massachusetts, 1964.
56. Rishel, R. W., Weak solutions of a partial differential equation of dynamic programming, *SIAM J. Control* **9**, No. 4 (1971), 519–528.
57. Rishel, R. W., Dynamic programming and minimium principles for systems with jump Markov disturbances, *SIAM J. Control* **13**, No. 2 (1975), 338–371.
58. Rishel, R. W., Optimality of controls for systems with jump Markov distrubances, in "Techniques of Optimization" (A. V. Balakrishnan, ed.) Academic Press, New York, 1972, pp. 335–352.
59. Rosonoer, L. I., L. S. Pontryagin's maximum principle in optimal system theory, II, *Automat. Remote Control* **20** (1959), 10–12.
60. Skorokhod, A. V., "Studies in the Theory of Random Processes." Addison-Wesley, Reading, Massachusetts, 1965, pp. 42–96.
61. Sworder, D. D., Feedback control of a calss of linear stochastic systems, *Proc. Joint Automatic Control Confer.* (1968), 34–44.
62. Sworder, D. D., Feedback control of a class of linear systems with jump parameters, *IEEE Trans. Automatic Control* **AC-14**, No. 1 (1969), 9–14.

63. Sworder, D. D., On the control of stochastic systems, *Internat. J. Control* **6** (1967), 179–188.
64. Sworder, D. D., On the stochastic maximium principle, *J. Math. Anal. Appl.* **24** (1968), 627–640.
65. Stroock, D. W., and Varadhan, S. R. S., Diffusion processes with continuous coefficients, I, *Comm. Pure Appl. Math.* **22** (1967), 345–400.
66. Teo, K. L., Optimal Control of Systems Governed by Ito Stochastic Differential Equations, Ph.D. thesis, Dept. Electr. Engrg., Univ. of Ottawa, 1974
67. Teo, K. L., and Ahmed, N. U., Optimal feedback control for a class of stochastic systems, *Internat. J. System Sci.* **5** (1974), 357–365.
68. Teo, K. L., Ahmed, N. U., and Wong, H. W., On optimal parameter selection for parabolic differential systems, *IEEE Trans. Automatic Control* **AC-19**, No. 3 (1964), 286–287.
69. Varaiya, P., Optimal control of a partially observed stochastic system, *Proc. Symp. Appl. Math., New York* (1972), 173–187.
70. Warga, J., Functions of relaxed controls, *SIAM J. Control* **5** (1967), 628–641.
71. Warga, J., "Optimal Control of Differential and Functional Equations." Academic Press, New York, 1972, pp. 346–352.
72. Wonham, W. M., On the separation theorem of stochastic control, *SIAM J. Control* **6** (1968), 321–326.
73. Wonham, W. M., Optimal stochastic control, *Automatica* **5** (1969), 113–118.
74. Wonham, W. M., Random differential equations in control theory, *in* "Probabilistic Methods in Applied Mathematics" (A. T. Bharucha-Reid, ed.), Vol. 2. Academic Press, New York, 1970, pp. 131–212.
75. Yong, L. C., Lectures on Calculus of Variations and on Optimal Control," Lecture Notes. Saunders, Philadelphia, Pennsylvania, 1969.

AMS (MOS) 1980 Subject Classifications: 93E20, 60H10

Gleason Measures

RYSZARD JAJTE*

DEPARTMENT OF MATHEMATICS
WAYNE STATE UNIVERSITY
DETROIT, MICHIGAN

I.	Introduction	69
II.	Generalities	71
III.	Orthogonally Scattered Gleason Measures	75
IV.	$L_{2,\xi}(H)$-Spaces and Isometries Generated by OSG Measures	80
V.	Spectral Gleason Measures	82
VI.	Convergence of Gleason Measures	89
VII.	Gleason Measures in Tensor Products	94
VIII.	Random Gleason Measures	97
	References	103

I. Introduction

The main purpose of this work is to give a rather systematic and self-contained survey of some recent results in the theory of measures on the lattice of all closed subspaces of a Hilbert space. At the same time we clarify the ideas and simplify some proofs. Many results seem to be presented here for the first time.

The entire theory discussed in this paper depends heavily on the classical result of Gleason [9, 42] giving the general form of a positive finite measure defined on the lattice Proj H of all orthogonal projectors acting in a separable

* On leave from Łódź University, Poland, while this research was being carried out.

Hilbert space H ($\dim H > 2$). Namely, for every such measure μ we have the following representation:

$$\mu P = \operatorname{tr} MP, \qquad P \in \operatorname{Proj} H, \tag{1.1}$$

where M is a linear, positive, self-adjoint, trace-class operator acting in H.

We shall be concerned with some vector- and operator-valued measures of Gleason type (i.e., defined on the lattice of all orthogonal projectors in a Hilbert space). Such measures can be extended to be linear operators defined on the algebra $L(H)$ of all bounded linear operators acting in H, so from the point of view of functional analysis, the theory of Gleason measures is in fact the theory of a special class of linear operators on $L(H)$. The general theory of linear mappings defined on von Neumann algebras [6] and C^*-algebras has been developed extensively during last years (e.g., [7, 21, 32, 36, 38, 40, 41]). However, Gleason's theorem is so deep and powerful that it is worth studying all its consequences. That is why we shall confine ourselves to the special von Neumann algebra $L(H)$ of all bounded linear operators in H. It should be noticed here that the Gleason measures are closely related to the description of quantum mechanical systems and the theory of the measures defined on the lattice of orthogonal projectors has as its motivation the mathematical foundations of quantum mechanics.

In quantum mechanics, to any physical system there corresponds a Hilbert space (cf. [23, 26]). The physical quantities (observables) are described by self-adjoint operators. If μ is a probability Gleason measure (state) and $A = \int \lambda E_A(d\lambda)$ is a spectral decomposition of an observable A, then the Borel measure p on the real line $\mathbb{R}$ defined by the formula $p(\cdot) = \mu(E_A(\cdot))$ is interpreted as a probability distribution of values of the observable A in the state μ. The evolution of a quantum mechanical system may be described either by changes of the states (Schroedinger picture) or by changes of the observables (Heisenberg picture). In both cases the one-parameter group of automorphisms of the logic $\operatorname{Proj} H$ and its action on states (or observables) is considered. The well-known theorem of Wigner [42, 43] says that any automorphism of $\operatorname{Proj} H$ is of the form $P \to UPU^{-1}$ for $P \in \operatorname{Proj} H$, where U is an arbitrary unitary or antiunitary operator. The Wigner automorphism may be treated as an operator-valued measure on $\operatorname{Proj} H$. It is a very special case of the spectral Gleason measures considered in Section V. Some general facts concerning Banach space-valued Gleason measures are discussed in Section II. Sections III and IV are devoted to orthogonally scattered Gleason measures. In Section V, using the results of Section III, we shall deal with an important class of projector-valued Gleason measures. This class seems to be closely related to quantum mechanics (mainly because of the spectral representation of observables). In Section VI we discuss some types of convergence in spaces of Gleason measures, and in Section VII the problems of extension of "cylindrical" Gleason measures in tensor products of Hilbert spaces are dealt with. The last section, Section VIII, is devoted to random Gleason measures.

II. Generalities

2.1 Let us begin with some notation. Let H, K be separable, complex Hilbert spaces, $\dim H > 2$. Let $\operatorname{Proj} H$ denote the lattice of all orthogonal projectors acting in H and let $L(H)$ be the algebra of all bounded linear operators in H. A nonnegative, self-adjoint, trace-class operator will be called for short an s-operator. For a unit vector $e \in H$, the symbol $[e]$ denotes the orthogonal projector on the subspace spanned by e, i.e., $[e]: y \to (y,e)e$ for all $y \in H$. E will denote a complex Banach space. $[x, y, \ldots]$ denotes the closed linear subspace spanned by the vectors $x, y, \ldots$.

2.2 Definition A mapping $\xi: \operatorname{Proj} H \to E$ is said to be an E-valued Gleason measure iff

(i) for any sequence of pairwise orthogonal projectors $P_1, P_2, \ldots$ from $\operatorname{Proj} H$

$$\sum_i \xi P_i = \xi\left(\sum_i P_i\right), \tag{2.1}$$

the series on the left-hand side being weakly convergent,

(ii) $\sup\{\|\xi P\| : P \in \operatorname{Proj} H\} < \infty$.

By the well-known theorem of Orlicz [29] the accepted definition implies the unconditional and strong convergence of (2.1).

A Gleason measure $\xi: \operatorname{Proj} H \to E$ will be called basic (for E) if the closed linear subspace of E spanned by the values of ξ coincides with E.

2.3 Recently, Sherstnev [39] gave the following generalization of Gleason's theorem (see Introduction) for bounded real-valued Gleason measures.

2.4 Theorem [39] *Let $\xi: \operatorname{Proj} H \to \mathbb{R}$ be a real-valued Gleason measure. Then there exists a self-adjoint, nuclear operator N such that*

$$\xi P = \operatorname{tr} NP \quad \text{for} \quad P \in \operatorname{Proj} H. \tag{2.2}$$

Proof Sherstnev's proof of this theorem depends heavily on some constructions used in the proof of the Gleason theorem. The result of Sherstnev can be easily deduced from the Gleason theorem. Indeed, let Z be a finite-dimensional subspace of H. Then the function α defined for $P \in \operatorname{Proj} Z$ by the formula

$$\alpha: P \to c \operatorname{tr} P + \xi P, \tag{2.3}$$

where $c = \sup\{|\xi[e]| : e$ runs over a unit sphere in $H\}$, is a positive Gleason measure. Hence, by the Gleason theorem

$$\alpha P = \operatorname{tr} M_Z P, \qquad P \in \operatorname{Proj} Z, \qquad (2.4)$$

for some self-adjoint operator M_Z in Z. Putting $N_Z = M_Z - cI_Z$ (I_Z the identity on Z), we obtain

$$\xi P = \operatorname{tr} N_Z P, \qquad P \in \operatorname{Proj} Z. \qquad (2.5)$$

Let us remark now that for two subspaces $Z_1 \subset Z_2$, the operators N_{Z_1}, N_{Z_2} are consistent (as determined by ξ). Now, for $x, y \in H$ let us put

$$a(x, y) = (N_Z, x, y), \qquad (2.6)$$

where Z is the subspace spanned by vectors x and y. $a(x, y)$ is then uniquely defined and homogeneous. Let now Y be the space spanned by vectors $x, x', y \in H$. Then

$$a(x + x', y) = (N_Y(x + x'), y) = (N_Y x, y) + (N_Y x', y)$$
$$= a(x, y) + a(x', y). \qquad (2.7)$$

Moreover,

$$\sup_{\|x\|=1} |a(x, x)| \leqslant \sup_{\|x\|=1} |\xi[x]| < \infty. \qquad (2.8)$$

Thus, there exists a bounded, linear operator N such that $a(x, y) = (Nx, y)$ for all $x, y \in H$. It is easy to check that N is self-adjoint and nuclear, which ends the proof.

Of course, the theorem of Sherstnev can be formulated also for complex-valued Gleason measures.

2.5 Corollary *Let $\xi: \operatorname{Proj} H \to E$ be a Banach space-valued Gleason measure. Then for any $x^* \in E$ there exists a trace-class operator $M(x^*)$ such that*

$$\langle \xi P, x^* \rangle = \operatorname{tr} M(x^*) P, \qquad P \in \operatorname{Proj} H, \qquad (2.9)$$

or, more exactly, there exist two (uniquely defined) self-adjoint trace-class operators $M_1(x^*), M_2(x^*)$ such that

$$\langle \xi P, x^* \rangle = \operatorname{tr} M_1(x^*) P + i \operatorname{tr} M_2(x^*) P = \operatorname{tr} M(x^*) P. \qquad (2.9')$$

2.6 Let us denote by $C_1(H)$ the Banach space of all trace-class operators acting in H with the norm

$$\|M\|_1 = \operatorname{tr}(M^*M)^{1/2} \qquad (2.10)$$

(cf., for example [35]).

By Corollary 2.5, with every E-valued Gleason measure ξ we can associate a family $\{M(x^*): x^* \in E^*\} \subset C_1(H)$, where the operators $M(x^*)$ are given by (2.9).

2.7 Theorem [11] Let $\xi: \text{Proj } H \to E$ be a Gleason measure. Then there exists a unique linear continuous operation

$$M: E^* \to C_1(H) \qquad (2.11)$$

such that (2.9) holds [E^* and $C_1(H)$ are endowed with norm topologies]. If the Banach space E is reflexive, then for any continuous operation (2.11) there exists a Gleason measure $\xi: \text{Proj } H \to E$ uniquely determined and satisfying (2.9).

Proof The linearity is obvious, so it suffices to prove the continuity. We shall show first that

$$\|M\|_1 \leq 4 \sup\{|\text{tr } MP|: P \in \text{Proj } H\}. \qquad (2.12)$$

Indeed, let $M \in C_1(H)$ and $M = M^*$. Then there exist two projectors P, $Q \in \text{Proj } H$ such that

$$\|M\|_1 = \text{tr } MP - \text{tr } MQ. \qquad (2.13)$$

To show it we write M in a diagonal form

$$M = \sum_k \lambda_k [e_k], \qquad (2.14)$$

where $\{e_K\}$ is an orthonormal basis, λ_k are real, and $\sum_k |\lambda_k| = \text{tr}(M^*M)^{1/2} = \|M\|_1 < \infty$. It suffices to put

$$P = \sum_{\lambda_k \geq 0} [e_K], \quad Q = \sum_{\lambda_k \geq 0} [e_K]. \qquad (2.15)$$

Now let M be an arbitrary operator from $C_1(H)$. Writing $M = M_1 + iM_2$, where M_j ($j = 1, 2$) are self-adjoint, we will obtain (2.12). Then the estimation

$$\|M(x^*)\| \leq 4 \sup_{P \in \text{Proj } H} |\text{tr } M(x^*)P| = 4 \sup|\langle \xi P, x^* \rangle| \leq 4C\|x^*\| \qquad (2.16)$$

yields the continuity.

On the other hand, let M be a linear continuous operation from E^* into $C_1(H)$. For every $P \in \text{Proj } H$ and $x^* \in E^*$ we have $M(x^*)P \in C_1(H)$ and the functional

$$x^* \to \text{tr } M(x^*)P$$

is linear and bounded. In fact,

$$|\text{tr } M(x^*)P| \leq \|M(x^*)P\|_1 \leq \|M(x^*)\|_1 \cdot \|P\|_\infty \leq \|M\| \cdot \|x^*\| \qquad (2.17)$$

(cf. [35, Theorem 2.3.10, p. 93]). Thus there exists a vector $\xi P \in E$ (E is reflexive) such that

$$\langle \xi P, x^* \rangle = \operatorname{tr} M(x^*)P.$$

The mapping $\xi \colon \operatorname{Proj} H \to E$ is a Gleason measure, because

$$\sup\{\|\xi P\| \colon P \in \operatorname{Proj} H\} = \sup\{|\langle \xi P, x^* \rangle| \colon P \in \operatorname{Proj} H, \|x^*\| = 1\}$$
$$\leq \|\|M\|\| < \infty.$$

2.8 Theorem *Every Gleason measure $\xi \colon \operatorname{Proj} H \to E$ can be extended in a unique way to a continuous linear operator $\xi \colon L(H) \to E$ [$L(H)$ and E are endowed with the uniform and strong topologies, respectively].*

Proof For a "simple operator"

$$A = \sum_{k=1}^{n} \lambda_k P_k, \tag{2.18}$$

where $\lambda_k \in \mathbb{C}$, $P_k \in \operatorname{Proj} H$, we put

$$\xi A = \sum_k \lambda_k \xi P_k. \tag{2.19}$$

Then, by Corollary 2.5, ξA is well defined and linear for simple operators. Because simple operators lie densely in $L(H)$ (in the uniform topology), it is sufficient to show the boundedness of ξ for simple operators. We shall prove that

$$\|\xi A\| \leq 4 \sup\{\|\xi P\| \colon P \in \operatorname{Proj} H\} \cdot \|A\| \tag{2.20}$$

for any simple operator A.

Let us fix an arbitrary simple operator A. We can write

$$\|\xi A\| = \langle \xi A, x_A^* \rangle \tag{2.21}$$

for some $x_A^* \in X^*$, $\|x_A^*\| = 1$ and then, by Corollary 2.5,

$$\langle \xi P, x_A^* \rangle = \operatorname{tr} M_1 P + i \operatorname{tr} M_2 P, \quad P \in \operatorname{Proj} H, \tag{2.22}$$

for some self-adjoint trace-class operators M_i. Let P_i^+, P_i^- ($i = 1, 2$) be orthogonal projectors such that

$$|M_i| = P_i^+ M_i - P_i^- M_i \quad (i = 1, 2) \tag{2.23}$$

(cf. the proof of Theorem 2.7). Then, by (2.22) and (2.23),

$$\|\xi A\| \leq \|A\|(\operatorname{tr}|M_1| + \operatorname{tr}|M_2|)$$
$$\leq \|A\|(\|\xi P_1^+\| + \|\xi P_2^+\| + \|\xi P_1^-\| + \|\xi P_2^-\|). \tag{2.24}$$

This gives (2.20) and concludes the proof of the theorem.

III. Orthogonally Scattered Gleason Measures

We shall now single out an important class of vector-valued Gleason measures: orthogonally scattered measures. The theory of orthogonally scattered measures on δ-rings of sets has been founded and developed by Masani [24, 24a]. The theory of orthogonally scattered Gleason measures can be treated as the extension of Masani's ideas to the lattice of projections in a Hilbert space.

Let H, K be two separable Hilbert spaces, $\dim H > 2$.

3.1 Definition A Gleason measure $\xi: \operatorname{Proj} H \to K$ is said to be orthogonally scattered (OSG-measure) if $P \perp Q$ implies $\xi P \perp \xi Q$ for any $P, Q \in \operatorname{Proj} H$.

3.2 Remark In the definition of OSG-measure we can omit the condition (ii) of Definition 2.1 because in this case $\|\xi P\| \leq \|\xi I\|$ for all P.

3.3 For any OSG-measure ξ the formula

$$\mu P = \|\xi P\|^2, \qquad P \in \operatorname{Proj} H, \tag{3.1}$$

defines some positive Gleason measure and by the Gleason theorem there exists an s-operator M such that

$$\|\xi P\|^2 = \operatorname{tr} MP, \qquad P \in \operatorname{Proj} H, \tag{3.2}$$

holds. Evidently,

$$(\xi P, \xi Q) = \operatorname{tr} MPQ \tag{3.3}$$

for any commuting projectors $P, Q \in \operatorname{Proj} H$.

The following theorem gives the general form of the "correlation function" $\{P, Q\} \to (\xi P, \xi Q)$ of OSG-measures.

3.4 Theorem [19] *Let H, K be complex, separable Hilbert spaces, $\dim H > 2$. If $\xi: \operatorname{Proj} H \to K$ is an OSG-measure then there exist two s-operators (i.e., positive and trace-class) M and N acting in H such that*

$$(\xi P, \xi Q) = \operatorname{tr} MPQ + \operatorname{tr} NQP, \qquad P, Q \in \operatorname{Proj} H. \tag{3.4}$$

When $\dim H = \infty$, then the operators M and N are uniquely determined by ξ.

For the sake of completeness and because of the importance of this theorem we sketch here the main steps of the proof. The following two lemmas are crucial for the proof of our theorem.

3.5 Lemma *Let H_n be the n-dimensional complex Hilbert space with an orthonormal basis $e_1, e_2, \ldots, e_n$, and a matrix $(T_{pqrs})_{p,q,r,s=1,\ldots,n}$ that satisfies* *the conditions*

$$T_{pqrs} = \overline{T_{rspq}}, \qquad p,q,r,s = 1,\ldots,n, \tag{3.5}$$

$$\sum_{p,q,r,s=1}^{n} T_{pqrs}(Pe_p, e_q)(Qe_r, e_s) = 0 \tag{3.6}$$

for any commuting operators $P, Q \in \text{Proj } H$. *Then*

$$T_{abcd} = 0 \quad \text{if } a \neq c \text{ and } b \neq d, \tag{3.7}$$

$$T_{atbt} = \overline{T_{btat}} = -T_{tbta} = -T_{tatb} = \mu_{ab} \tag{3.8}$$

if $a \neq b, t = 1, \ldots, n$ (*i.e.*, T_{atbt} *does not depend on t*), *and*

$$T_{abab} = -T_{baba} \quad (\text{in particular, } T_{aaaa} = 0), \tag{3.9}$$

$$T_{abab} + T_{bcbc} + T_{caca} = 0 \tag{3.10}$$

for any $a,b,c,d = 1,\ldots,n$.

Sketch of the Proof of the Lemma Putting in (3.6) $P = [e_a]$, $Q = [e_b]$, where $a, b = 1, 2, \ldots, n$, we obtain

$$T_{aabb} = 0, \qquad a,b = 1,2,\ldots,n. \tag{3.11}$$

Now if we put $P = Q = [\alpha e_a + \beta e_b]$, where α, β are complex numbers, $|\alpha|^2 + |\beta|^2 = 1$, $a \neq b$, and $a, b = 1, 2, \ldots, n$, then the polynomial obtained is homogeneous with respect to α, β. It is easy to check that the coefficients of all different products of the variables α, $\bar{\alpha}$, β, $\bar{\beta}$ must vanish, so we have

$$T_{abbb} = -T_{bbba}, \qquad T_{abaa} = -T_{aaba} \tag{3.12}$$

and

$$T_{abab} = -T_{baba}, \qquad T_{abba} = T_{baab} = 0. \tag{3.13}$$

Similarly, putting in (3.6)

$$p = [\alpha e_a + \beta e_b], \qquad Q = [e_a, e_b],$$

we obtain (as a result of the vanishing of the coefficient at $\bar{\alpha}\beta$)

$$T_{abbb} = -T_{abaa}.$$

Thus, by (3.12)

$$T_{abbb} = -T_{bbba} = -T_{abaa} = T_{aaba}$$

and by (3.5) we obtain (3.8) for $t \in \{a, b\}$. Note that (3.7) for $c, d \in \{a, b\}$ is reduced to

$$T_{aabb} = T_{bbaa} = T_{abba} = T_{baab} = 0.$$

Therefore, by (3.11) and (3.13), all the formulas (3.7)–(3.9) are satisfied in the case of the indices a, b, c, d, t taking at most two different values. The formula (3.8) is now a consequence of

$$T_{acbc} = -T_{cbca} = \mu_{ab}, \qquad a \neq b \neq c \neq d, \tag{3.14}$$

and, by (3.5), the condition

$$T_{ccab} = T_{abbc} = 0 \tag{3.15}$$

implies (3.7) when three of the indices a, b, c, d at the most may be mutually different.

If we put $P = [\alpha e_a + \beta e_b, e_c]$, $Q = [\delta(\alpha e_a + \beta e_b) + \gamma e_c]$, where $|\alpha|^2 + |\beta|^2 + |\gamma|^2 + |\delta|^2 = 1$, then (3.6) gives a homogeneous polynomial with respect to the variables γ, δ. The coefficient at $\delta\bar{\gamma}$ is a homogeneous polynomial with respect to α, β and as the coefficients at $\bar{\alpha}\beta^2$, $|\alpha|^2\beta$ and $\alpha|\beta|^2$, respectively, we get

$$T_{abbc} = 0 \tag{3.16}$$

$$T_{abac} + T_{aabc} + T_{ccbc} = 0, \qquad T_{bbac} + T_{babc} + T_{ccac} = 0. \tag{3.17}$$

Now we put

$$P = [\zeta(\alpha e_a + \beta e_b + e_c)],$$
$$Q = [\zeta'(\alpha' e_a + \beta' e_b - (\bar{\alpha}\alpha' + \bar{\beta}\beta')e_c)],$$

where

$$\zeta = (|\alpha|^2 + |\beta|^2 + 1)^{-1/2},$$
$$\zeta' = (|\alpha'|^2 + |\beta'|^2 + |\bar{\alpha}\alpha' + \bar{\beta}\beta'|^2)^{-1/2}, \quad |\alpha'| + |\beta'| > 0.$$

Then $P \perp Q$. The polynomial obtained is now extremely long, but we can easily write the coefficients at $\alpha'\bar{\beta}'$, $\bar{\alpha}\beta\alpha'\bar{\beta}'$ and obtain

$$T_{ccab} = 0, \tag{3.18}$$

$$T_{abab} - T_{cbcb} - T_{acac} + T_{cccc} = 0, \tag{3.19}$$

the last formulas needed to show (2.5)–(2.8) in the case $\dim H_n = n = 3$ and when indices a, b, c, d, t can take at most three different values. To prove the lemma it is enough to exhibit (3.7) for mutually different a, b, c, d. For this purpose we put into (3.6)

$$P = [\alpha e_a + \beta e_b], \qquad Q = [\gamma e_c + \delta e_d]$$

(where $|\alpha|^2 + |\beta|^2 = |\gamma|^2 + |\delta|^2 = 1$), and thus we obtain $T_{abcd} = 0$ (as the coefficient at $\bar{\alpha}\beta\gamma\bar{\delta}$) which ends the proof of our lemma.

Let us write the density operator M from the formula (3.2) in the diagonal form

$$M: e_j \to \lambda_j e_j, \quad j \in J, \quad (3.20)$$

(i.e., the sequence $\{e_j\}$ is an orthogonal basis in H of eigenvectors of M). Let us put $H_n = [e_1, e_2, \ldots, e_n]$, $n \in J$.

3.6 Lemma *There exists a matrix* $(U_{pqrs})_{p,q,r,s \in J}$ *such that for any* $u \in J$ *and any projectors* $P, Q \in \mathrm{Proj}\, H_n$ *we have*

$$(\xi P, \xi Q) = \sum_{pqrs=1}^{n} U_{pqrs}(Pe_p, e_q)\overline{(Qe_r, e_s)} \quad (3.21)$$

and

$$U_{pqrs} = \overline{U_{rspq}} \quad (3.22)$$

for any $p, q, r, s = 1, 2, \ldots, n$.

We omit the rather standard proof. The details can be found in Jajte and Paszkiewicz [19].

3.7 *Sketch of the Proof of Theorem* 3.3 Let the operator M be given by (3.2). For any commuting projectors $P, Q \in \mathrm{Proj}\, H_n$ we have by (3.3)

$$(\xi P, \xi Q) = \sum_{pqrs=1}^{n} \lambda_p \delta_{pr} \delta_{qs} (Pe_p, e_q) \overline{(Qe_r, e_s)}$$

(δ_{ab} is the Kronecker symbol).

Let us put

$$T_{pqrs} = U_{pqrs} - \lambda_p \delta_{pr} \delta_{qs}, \quad p, q, r, s = 1, \ldots, n.$$

Then, by Lemma 3.6, the formulas (3.5) and (3.6) are valid for any commuting operators $P, Q \in \mathrm{Proj}\, H_n$ and, consequently, by Lemma 3.5, the formulas (3.7)–(3.10) hold for any $p, q, r, s = 1, 2, \ldots, n$ and thus for any $p, q, r, s \in J$.

Let us put

$$\tilde{\alpha}_1 = 0, \quad \tilde{\alpha}_p = T_{1p1p}, \quad p \in J, \quad p \geq 2.$$

Then

$$T_{pqpq} = T_{1q1q} - T_{1p1p} = \tilde{\alpha}_q - \tilde{\alpha}_p, \quad p, q \in J.$$

Putting $\tilde{\beta}_p = \lambda_p - \tilde{\alpha}_p$, $p \in J$, we have

$$U_{pqpq} = T_{pqpq} + \lambda_p = \tilde{\beta}_p + \tilde{\alpha}_q.$$

Moreover (for $a, b \leq n \in J$),

$$\tilde{\beta}_a + \tilde{\alpha}_b = U_{abab} \geq 0.$$

Thus, for
$$\alpha_a = \tilde{\alpha}_a - \inf_{p \in J} \tilde{\alpha}_p, \qquad \beta_a = \lambda_a - \alpha_a,$$
we have
$$\alpha_a + \beta_a = \lambda_a, \qquad \beta_a + \alpha_b = U_{abab}, \qquad \alpha_a \geq 0, \quad \beta_a \geq 0. \tag{3.23}$$

One can verify that
$$U_{pqrs} = \mu_{pr}\delta_{qs} - \overline{\mu_{qs}}\delta_{pr} + (\beta_p + \alpha_q)\delta_{pr}\delta_{qs}. \tag{3.24}$$

Now, we define the matrices (for $a, b \in J$)
$$m'_{ab} = -\mu_{ab} + \delta_{ab}\alpha_a, \qquad m''_{ab} = \mu_{ab} + \delta_{ab}\alpha_a. \tag{3.25}$$

For any operators $A, B \in L(H_n)$ we now have
$$(\xi A, \xi B) = \sum_{p,q,r,s=1}^{n} U_{pqrs}(Ae_p, e_q)\overline{(Be_r, e_s)}$$
$$= \sum_{p,r=1}^{n} m''_{pr}(Ae_p, Be_r) + \sum_{q,s=1}^{n} m'_{qs}(Be_q, e_s). \tag{3.26}$$

One can prove [19] that the formulas
$$(e^a, Ne_b) = m'_{a,b}, \qquad (e_a, Me_b) = m''_{a,b}, \qquad a, b \in J, \tag{3.27}$$

define s-operators N and M in H such that
$$(\xi A, \xi B) = \operatorname{tr} NAB^* + \operatorname{tr} MB^*A \tag{3.28}$$

holds for all $A, B \in L(H_n)$ (after zero extension on the vectors e_p, $p \geq n$). The standard reasoning leads us to the general case $L(H)$. The representation (3.4) of the correlation function of ξ is unique when $\dim H = \infty$. In fact, let $\operatorname{tr} MB^*A + \operatorname{tr} NAB^* = 0$ and let us fix some point $x_0 \in H$, $\|x_0\| = 1$. Then putting $A = U[x_0]$, $B = U$, where U is an arbitrary unitary operator, we obtain $(Nz, z) = \text{const.}$ for all $z \in H$, $|z| = 1$. It is possible only when $N = 0$, because N is nuclear operator. Similarly, $M = 0$, which completes the proof.

3.8 Theorem 3.4 has been formulated and proved for the case of complex Hilbert spaces. This theorem is also true for real Hilbert spaces. In this case the formula (3.4) can be reduced to
$$(\xi P, \xi Q) = \operatorname{tr} MPQ = \operatorname{tr} MQP, \qquad P, Q \in \operatorname{Proj} H. \tag{3.29}$$

The proof of (3.29) for real spaces is very easy. Namely, (3.29) follows immediately from the formula
$$\|\xi(P + Q)\|^2 = \operatorname{tr} M(P + Q)^2, \qquad P, Q \in \operatorname{Proj} H, \tag{3.30}$$

where ξ in (3.30) denotes a natural extension of ξ onto $L(H)$ and M is given by Gleason's theorem

$$\|\xi P\|^2 = \operatorname{tr} MP, \qquad P \in \operatorname{Proj} H. \tag{3.31}$$

Let us remark that in the complex case the formula

$$\|\xi A\|^2 = \operatorname{tr} MAA^* \tag{3.32}$$

is valid only for normal operators and, in particular, it does not hold for $A = P + iQ$ when P and Q do not commute. That is why the proof for the complex case is so complicated.

IV. $L_{2,\xi}(H)$-Spaces and Isometries Generated by OSG Measures

4.1 Let $\xi: \operatorname{Proj} H \to K$ be an OSG-measure and M and N be the operators from the formula (3.4). Every OSG-measure can be extended to the linear operator on $L(H)$ into K (this extension will be denoted by the same letter ξ). For $A, B \in L(H)$ we put

$$\langle A, B \rangle_\xi = (\xi A, \xi B)_K = \operatorname{tr} MAB^* + \operatorname{tr} NB^*A \tag{4.1}$$

[cf. (3.28)], where M and N are the operators given by Theorem 3.4.

Let $L_{2,\xi}(H)$ denote the completion of $L(H)/\langle A, A \rangle_\xi = 0$ (with respect to $\langle \cdot, \cdot \rangle_\xi$).

The structure of the points of $L_{2,\xi}$ is rather complicated but it is of no interest to us. Actually, we are interested in a construction of some Hilbert space generated by an OSG-measure ξ. This construction gives us the possibility to investigate some properties of operations defined on a "decent" part of $L_{2,\xi}$ [as $L(H)$ or $\operatorname{Proj} H$].

The importance of $L_{2,\xi}$-spaces lies in the following fact which is an immediate consequence of (4.1) and the definition of the space $L_{2,\xi}$.

4.2 Theorem *The* OSG-*measure* $\xi: \operatorname{Proj} H \to K$ *can be extended to the isometry on* $L_{2,\xi}$ *onto* $X_\xi \subset K$. *This isometry will also be denoted by* ξ. *This is justified by the uniqueness of this extension. Of course, ξ is basic for K if and only if the above extension of ξ is an isometry on $L_{2,\xi}$ onto K (i.e., $X_\xi = K$).*

4.3 Definition A Hilbert space $\mathcal{H}$ is said to be algebraic iff

(i) there exists a Hilbert space H such that $L(H) \subset \mathcal{H}$,

(ii) $L(H)$ is a pre-Hilbert space with some inner product $\langle \cdot, \cdot \rangle$ such that $B^*A = 0 = AB^*$ implies $\langle A, B \rangle = 0$ for all $A, B \in L(H)$ and such that for any decreasing sequence $\{P_n\}$ of orthogonal projectors in H, $P_n \to 0$ (in the strong operator topology) implies $\langle P_n, P_n \rangle \to 0$,

(iii) $\mathcal{H}$ is a completion of $\{L(H), \langle \cdot, \cdot \rangle\}$ (with respect to the inner product $\langle \cdot, \cdot \rangle$). The space H will be called the kernel of $\mathcal{H}$, and denoted by ker $\mathcal{H}$.

4.4 Example Let $\mu: \operatorname{Proj} H \to \mathbb{R}^+$ be an arbitrary positive, full Gleason measure. Taking the extension of μ to the linear functional on $L(H)$, we put $\langle A, B \rangle = \mu(B^*A)$. The completion of $L(H)$ under this inner product is an algebraic Hilbert space.

Of course, every $L_{2,\xi}(H)$-space is an algebraic Hilbert space with the kernel H.

4.5 Theorem *The class of all algebraic Hilbert spaces coincides with the class of all $L_{2,\xi}$-spaces.*

Proof It is sufficient to show that every algebraic Hilbert space is some $L_{2,\xi}$-space. Let $\mathcal{H}$ be a space as in Definition 4.3. We define $\xi: \operatorname{Proj} H \to \mathcal{H}$ by the formula

$$\xi P = P, \qquad P \in \operatorname{Proj} H, \tag{4.2}$$

where P on the right-hand side is treated as an element of $\mathcal{H}$ (a representative of an equivalence class). By (ii) of Definition 4.3, ξ is an OSG-measure and for $A, B \in L(H)$ we have

$$\langle A, B \rangle_\xi = \langle \xi A, \xi B \rangle_\mathcal{H}, \tag{4.3}$$

which ends the proof.

The OSG-measure defined by (4.4) will be called an indicator measure. Of course, this measure is basic for $\mathcal{H}$.

4.6 Corollary [15] *Let $\langle \cdot, \cdot \rangle$ denote the inner product on the algebra $L(H)$ satisfying the conditions (ii) of Definition 4.3. Then there exist two s-operators M and N acting in H such that*

$$\langle A, B \rangle = \operatorname{tr} MAB^* + \operatorname{tr} NB^*A. \tag{4.4}$$

That is, the arbitrary inner product on $L(H)$, consistent with the natural orthogonality in $L(H)$, is given by two nuclear, positive operators.

4.7 For an OSG-measure $\xi: \operatorname{Proj} H \to K$ and a point $x \in K$, let us denote by $P_\xi x$ the orthogonal projection of x into the space $X_\xi \subset K$.

4.8 Proposition *Let (i) $\xi: \operatorname{Proj} H \to K$ be an OSG measure, (ii) and for any $x \in K$ and $P \in \operatorname{Proj} H$, let $v_x(P) = (x, \xi P)$; then for any $x \in K$ there exists an exactly one point $d_x \in L_{2,\xi}(H)$ such that*

(a) $\operatorname{Proj}_\xi x = \xi(d_x)$,
(b) $v_x(P) = (\xi d_x, \xi P)$ *for all $P \in \operatorname{Proj} H$.*

Proof (a) The existence of an exactly one point $d_x \in L_{2,\xi}(H)$ satisfying (a) follows immediately from Theorem 4.2. (b) The orthogonality of $x - P_\xi x$ to X_ξ and (a) entail (b).

4.9 Lemma *Let H, K be two Hilbert spaces. Let f be an isometry on H into K and g be an isometry on K into H, such that*

$$(fx, y) = (x, gy) \quad \text{for all} \quad x \in H, y \in K. \tag{4.5}$$

Then $fg = gf = I$ (identity), so $f = g^{-1}$ and f and g are isometries "onto."

Proof For every $x', x'' \in H$ we have

$$(gfx', x'') = (fx', fx'') = (x', x'')$$

and, similarly,

$$(fgy', y'') = (y', y'') \quad \text{for any} \quad y', y'' \in K,$$

which ends the proof.

4.10 Theorem *Let*

(i) $\mathscr{H}_i$, where $i = 1, 2$, be two algebraic Hilbert spaces with $\ker \mathscr{H}_i = H_i$,
(ii) $\xi: L(H_2) \to \mathscr{H}_1$ and $\eta: L(H_1) \to \mathscr{H}_2$ be two OSG-measures,
(iii) ξ be consistent with $\mathscr{H}_2$, i.e., $(\xi A, \xi B)_{\mathscr{H}_1} = (A, B)_{\mathscr{H}_2}$ for all $A, B \in L(H_2)$,
(iv) η be consistent with $\mathscr{H}_1$, i.e., $(\eta A, \eta B)_{\mathscr{H}_2} = (A, B)_{\mathscr{H}_1}$ for all $A, B \in L(H_1)$, and let
(v) $(\eta A_1, A_2)_{\mathscr{H}_2} = (A_1, \xi A_2)_{\mathscr{H}_1}$ for any $A_i \in L(H_i)$.

Then ξ is basic for $\mathscr{H}_1$ and η is basic for $\mathscr{H}_2$.

Proof This is an immediate consequence of Theorem 4.2 and Lemma 4.9.

Theorem 4.10 can be treated as a version of the Bochner–Masani duality theorem [2, 24, 24a] for Gleason measures. The proof via Lemma 4.9 seems to be the simplest also in the case of the Bochner and Masani theorems.

V. Spectral Gleason Measures

5.1 In this section we shall examine a class of operator-valued Gleason measures which seems to us the most important. This class is closely related to the orthogonally scattered Gleason measures and, on the other hand, is connected with Wigner's theorem (see Introduction). In particular, we shall give here the generalization of Wigner's theorem for homomorphisms of the lattice Proj H.

Let us begin with some definitions.

5.2 Definition Let H, K be two Hilbert spaces. A mapping $\xi: \operatorname{Proj} H \to L(H)$ is called an operator-valued OSG-measure if for an arbitrary sequence of pairwise orthogonal projectors $P_1, P_2, \ldots$ from $\operatorname{Proj} H$ the operators ξP_i are mutually orthogonal, i.e.,

$$(\xi P_i)^* \xi P_j = 0 \quad \text{for} \quad i \neq j \quad (i = 1, 2, \ldots) \tag{5.1}$$

and

$$\sum_i \xi P_i = \xi\left(\sum_i P_i\right), \tag{5.2}$$

the series on the left-hand side being convergent in the weak operator topology. By a theorem of Orlicz [29], the series (5.2) is then convergent in the strong operator topology.

The orthogonally scattered operator-valued Gleason measure is said to be a spectral Gleason measure iff it is normalized, i.e.,

$$\xi I_H = I_K \tag{5.3}$$

holds, where I denotes the identity operator.

5.3 Proposition If ξ is a spectral Gleason measure, then $\xi(\operatorname{Proj} H) \subset \operatorname{Proj} K$, ie, ξ is a projector-valued Gleason measure.

Proof By (5.1) we have for any $P \in \operatorname{Proj} H$

$$(\xi P)^*(I_K - \xi P) = (\xi P)^* \xi(I_H - P) = 0. \tag{5.4}$$

Thus

$$(\xi P)^* = (\xi P)^* \xi P. \tag{5.5}$$

But the last operator is self-adjoint and, consequently, so is ξP and $\xi P = (\xi P)^2$, which ends the proof.

5.4 Theorem [19, 31] Let H, K be two separable (real or complex) Hilbert spaces, $\dim H > 2$. For any spectral Gleason measure $\xi: \operatorname{Proj} H \to \operatorname{Proj} K$ there exists a sequence $U_i: H \to K_i$ ($i \in I$) of isometric or antiisometric (onto K_i) operators such that K is an orthogonal sum $K = \bigoplus_{i \in I} K_i$, and

$$\xi P = \bigoplus_{i \in I} U_i P U_i^{-1} \tag{5.6}$$

for any $P \in \operatorname{Proj} H$. In the other words, every spectral Gleason measure can be decomposed into an orthogonal sum of Wigner's isomorphisms. Evidently, in case of real Hilbert spaces, all the operators U_i in the formula (5.6) are isometric.

The card(I) will be called the dimension of ξ and denoted by $\dim \xi$. Of course, $\dim \xi = \dim \xi[e]$ for any unit vector e. If the spectral measure $\xi : \operatorname{Proj} H \to \operatorname{Proj} K$ is an isomorphism, then $\dim \xi = 1$ and we obtain the theorem of Wigner.

Sketch of the Proof We shall prove the theorem only for real Hilbert spaces [19]. In the complex case the proof is longer and more complicated but the general idea is similar (cf. [31]).

In the sequel we shall identify the orthogonal projector P with the subspace on which it projects. For $a \neq 0$,
$$[a] : y \to (y, a/\|a\|)(a/\|a\|).$$

Let us fix some unit vector $e \in H$. For any vector $x \in K$ the formula
$$\xi_x P = (\xi P)x, \qquad P \in \operatorname{Proj} H, \tag{5.7}$$
defines an OSG-measure and, by (3.29), there exists an s-operator M_x such that
$$(\xi_x P, \xi_x Q) = \operatorname{tr} M_x PQ. \tag{5.8}$$
Let us assume now that $x \in \xi[e]$. Then
$$\operatorname{tr} M_x[e] = \|\xi[e]x\|^2 = \|x\|^2 = \operatorname{tr} M_x; \tag{5.9}$$
thus
$$M_x = \|x\|^2 [e], \tag{5.10}$$
and for vectors $a, b \in H$, $\|a\| = \|b\| = 1$, we obtain
$$(\xi[a]x, \xi[b]x) = \|x\|^2 \operatorname{tr}[e][a][b] = \|x\|^2 (e, a)(e, b)(a, b). \tag{5.11}$$
For $x \in \xi[e]$ and $a \in H$, $(a, e) \neq 0$, we define a function
$$U_e(x, a) = [\|a\|^2/(a, e)]\xi[a]x. \tag{5.12}$$
Then we have
$$U_e(x, e) = x, \tag{5.13}$$
$$(U_e(x, a), U_e(x, b)) = \|x\|^2 (a, b) \tag{5.14}$$
(for $(a, e) \neq 0 \neq (b, e)$, $x \in \xi[e]$).

For a fixed x, $\|x\| = 1$, the function $U_e(x, \cdot)$ can be extended to be the isometric operator on H into K. The operators $U_e(f, \cdot)$, where $\|f\| = 1$, $f \in \xi[e]$, have the following properties:

(α) $U_e(f, a) \in \xi[a]$, $a \in H$, $a \neq 0$;
(β) $f \perp f'$, $f, f' \in \xi[a]$ imply $U_e(f', b)$, $a, b \in H$;
(γ) if $f' = U_e(f, e')$, $e' \in H$, $\|e'\| = 1$, then $U_{e'}(f', \cdot) = U_e(f, \cdot)$

Property (α), for $(a, e) \neq 0$, immediately follows from (5.12). If $(a, e) = 0$, $a \neq 0$, we put $a_n = a + (1/n)e$ and $e_n = a_n/\|a_n\|$ $(n = 1, 2, \ldots)$. Then
$$f_n = U_e(f, a_n) \in \xi[a_n], \qquad U_{e_n}(f^n, a/\|a\|) \in \xi[a],$$
and we have
$$\|U_e(f, a) - U_{e_n}(f_n, a/\|a\|)\| \leq \|U_e(f, a) - f_n\| + \|f_n - U_{e_n}(f_n, a/\|a\|)\|$$
$$= \|U_e(f, a - a_n)\|$$
$$+ \|U_{e_n}(f_n, e_n - (a/\|a_n\|))\| \to 0 \qquad \text{as} \quad n \to \infty,$$
so $U_e(f, a) = \lim U_{e_n}(f_n, a) \in \xi[a]$.

To prove (β) let us notice that
$$(U_e(f, a), U_e(f', a)) = \tfrac{1}{2}\{\|U^e(f + f', a)\|^2 - \|U_e(f, a)\|^2 - \|U_e(f', a)\|^2$$
$$= (\|a\|^2/2)(\|f + f'\|^2 - \|f\|^2 - \|f'\|^2)$$
$$= 0 \qquad \text{for any} \quad a \in H.$$

Let now $\{e_i\}$ be an orthonormal basis in H. Then $U_e(f, e_i) \perp U_e(f', e_i)$ and, by (α), $U_e(f, e_i) \perp U_e(f', e_j)$ for $i \neq j$. Therefore $U_e(f, a) \perp U_e(f', b)$ for all $a, b \in H$.

To prove (γ), let us put (for $(a, e) \neq 0 \neq (a, e')$]
$$x = f = U_e(f, e), \qquad y = f' = U_e(f, e') = U_{e'}(f', e'), \qquad P = \xi[a].$$
Then
$$Px = [(a, e)/\|a\|^2]U_e(f, a), \qquad Py = [(a, e')/\|a\|^2]U_{e'}(f', a),$$
$$\|Px\| = |(a, e)|/\|a\|, \qquad \|Py\| = |(a, e')|/\|a\|,$$
$$(Px, y) = (a, e)(a, e')/\|a\|^2.$$

Thus $(Px, Py) = (Px, y) = \operatorname{sgn}((a, e)(a, e'))\|Px\|\|Py\|$, which implies
$$Px = [(a, e)/(a, e')]Py$$
and $U_e(f, a) = U_{e'}(f', a)$. The last condition holds also if the vector a is orthogonal to e or e'. The property (γ) is proved.

Let us fix now an arbitrary orthonormal basis in $\xi[e]$ (for some unit vector $e \in H$). Using the properties (α)–(γ) it is not too difficult to show that the unitary operators
$$U_i: h \to U_e(f_i, h) \qquad (i = 1, 2, \ldots) \tag{5.15}$$
give the representation (5.6), which ends the proof of the theorem.

5.5 A spectral measure $\xi: \operatorname{Proj} H \to \operatorname{Proj} K$ can be extended to be a linear operation $\xi: L(H) \to L(K)$. Let H, K be complex spaces and let us rewrite the formula (5.6) in the form
$$\xi P = \bigoplus_i U_i P U^{-1} \oplus \bigoplus_j V_j P V_j^{-1}, \tag{5.16}$$

where the U_i are isometric and the V_j are antiisometric operators. When (5.16) reduces to the "unitary part," i.e.,

$$\xi P = \bigoplus_i U_i P U_i, \tag{5.17}$$

then the extension of ξ to the linear operator is given by the formula

$$\xi A = \bigoplus_i U_i P U_i^{-1}, \tag{5.18}$$

and it is a *-representation of the algebra $L(H)$ in K. In the general case (5.16), ξ cannot be extended to be a *-representation because, for $A = \sum_k \lambda_k P_k$, $P_k \in \text{Proj } H$, $\lambda_k \in \mathbb{C}$, we have

$$\xi(A) = \sum_k \lambda_k \xi(P_k) = \left(\bigoplus_i U_i A U_i^{-1} \right) \oplus \left(\bigoplus_j V_j A^* V_j^{-1} \right) \tag{5.19}$$

and, consequently,

$$\xi(AB) = \left(\bigoplus_i U_i A B U_i^{-1} \right) \oplus \left(\bigoplus_j V_j B^* A^* V_j^{-1} \right)$$

$$= (\xi_U A \xi_U B) \oplus (\xi_V B \xi_V A), \tag{5.20}$$

where

$$\xi_U(\cdot) = \bigoplus_i U_i(\cdot) U_i^{-1}, \qquad \xi_V = \bigoplus_j V_j(\cdot) V_j^{-1}. \tag{5.21}$$

Moreover, we have

$$\xi(A^*) = (\xi A)^* \qquad \text{for all} \quad A \in L(H). \tag{5.22}$$

Let us introduce the following definition.

5.6 Definition Let $\mathscr{A}$ and $\mathscr{B}$ be two C^*-algebras. A linear mapping $\eta: \mathscr{A} \to \mathscr{B}$ will be called a *-quasi-homomorphism of $\mathscr{A}$ into $\mathscr{B}$ iff
  (1) $\eta(a^*) = \eta(a)^*$,
  (2) $\eta(a \circ b) = \eta a \circ \eta b$,

where $a \circ b$ denotes a symmetrized product, i.e., $a \circ b = ab + ba$. When $\mathscr{B}$ is an algebra $L(K)$, then we say that η is a *-quasi-representation of $\mathscr{A}$ in K. A *-quasi-representation $\eta: L(H) \to L(K)$ is called normalized if $\eta(L_H) = I_K$ and η is called weakly continuous if it is a continuous linear mapping on $L(H)$ into $L(K)$ when $L(H)$ and $L(K)$ are endowed with the weak operator topologies.

It follows immediately from (5.20) and (5.22) that every spectral Gleason measure ξ can be extended to a normalized, weakly continuous *-quasi-representation of $L(H)$ in K. In fact the converse assertion is also true and we have the following theorem.

5.7 Theorem *The class of all normalized, weakly continuous *-quasi-representations of $L(H)$ in K coincides with the class of all spectral Gleason measures $\xi: L(H) \to L(K)$.*

Proof It is easy to check that $\xi(\operatorname{Proj} H) \subset \operatorname{Proj} K$ and $\xi P \perp \xi(I - P) = I - \xi P$, so it suffices to show that ξ is monotone, i.e.,

$$P \leq Q \quad \text{implies} \quad \xi P \leq \xi Q \quad \text{for } P, Q \in \operatorname{Proj} H. \tag{5.23}$$

Let us assume the contrary and, for some $P, Q \in \operatorname{Proj} H$, let $P \leq Q$ and $\xi P \not\leq \xi Q$ hold. That means that there exists an $x \in \xi P(K) \setminus \xi Q(K)$. Since $P \leq Q$, we have

$$\xi P \xi Q + \xi Q \xi P = \xi(P \circ Q) = 2\xi P$$

and, because $x = \xi P_x$, $\|x\| > \|\xi Q_x\|$, we would have

$$2\|x\| = 2\|\xi P x\| \leq \|\xi P \xi Q x\| + \|\xi Q \xi P x\| \leq 2\|\xi Q x\| < 2\|x\|,$$

which is impossible, so (5.23) holds. This concludes the proof of the theorem.

5.8 Definition Let $L^+(H)$ denote the positive cone in $L(H)$ of all symmetric positive operators. A mapping $\xi: \operatorname{Proj} H \to L^+(K)$ is said to be a positive (semispectral) Gleason measure when for any sequence $P_1, P_2, \ldots$ of mutually orthogonal projectors from $\operatorname{Proj} H$, the series $\sum_i \xi P_i$ converges in a weak operator topology to $\xi(\sum_i P_i)$ and $\xi(I_H) = I_K$.

By the Gleason theorem there is a unique extension of semi spectral measure to a linear operator $\xi: L(H) \to L(K)$.

In the commutative case, that is, when we consider measures on a σ-field of measurable sets, we have the well-known dilation theorem of Naimark [8, 25]. In the case of Gleason measures the dilation theorem is true only for a subclass of positive measures. That subclass has been described by Stinespring [40] in a general case of C*-algebras. Namely, let A and B be C*-algebras and ξ a positive map of A into B. The notion of a completely positive mapping used by Stinespring [40] is slightly different but equivalent to the notion of a positive definite mapping in case of an algebra B acting in a Hilbert space. That is why we use the condition of positive definiteness in the formulation of the following result of Stinespring.

5.9 Theorem [40] *Let ξ be a normalized positive linear map of a C*-algebra into another acting on a Hilbert space K. Then ξ is positive definite on the multiplicative semigroup A if and only if there exist a Hilbert space H, a linear isometry V of K into H, and a *-representation Π of A on H such that*

$$\xi a = V^* \Pi(a) V \quad \text{for all} \quad a \in A. \tag{5.24}$$

5.10 Let H, K be two Hilbert spaces. We say that an orthogonally scattered Gleason measure $\xi: \operatorname{Proj} H \to K$ originates from a spectral Gleason measure $\eta: \operatorname{Proj} H \to \operatorname{Proj} K$ when there exists a vector $k_0 \in K$ such that

$$\xi P = (\eta P) k_0 \quad \text{for all} \quad P \in \operatorname{Proj} H. \tag{5.25}$$

Then, of course, $k_0 = \xi I_H$. The description of the class of all OSG-measures $\xi: \operatorname{Proj} H \to K$ which originate from some Gleason spectral measures (i.e., from some *-quasi-representation) seems to be open. We shall give the characterization of such OSG-measures which originate from the *-representations.

5.11 Theorem *An orthogonally scattered Gleason measure $\xi: \operatorname{Proj} H \to K$ originates from a *-representation η of $L(H)$ in K if and only if the formula*

$$(\xi P, \xi Q) = (\xi(QP), \xi I) \tag{5.26}$$

holds for all $P, Q \in \operatorname{Proj} H$ (cf. [16]).

Proof Sufficiency: To avoid some trivial complications we assume that ξ is basic, i.e., the closed linear subspace spanned by the values of ξ coincides with K. The formula (5.26) can be rewritten for $a, b \in L(H)$ in the form

$$(\xi a, \xi b) = (\xi((b^*a), \xi I). \tag{5.27}$$

For any $A \in L(H)$ we define a bounded linear operator $\tilde{A} \in L(K)$, putting

$$\tilde{A}: \xi a \to \xi(Aa). \tag{5.28}$$

The above definition of $\tilde{A}$ is correct as $\xi a = 0$ implies $\xi(Aa) = 0$. In fact, if A is a unitary operator, then by (5.28) we have

$$\|\xi(Aa)\|^2 = (\xi(Aa), \xi(Aa)) = \xi(a^*A^*Aa), \xi I) = (\xi a, \xi a) = \|\xi a\|^2.$$

For an arbitrary $A \in L(H)$ we can choose unitary operators $U_1, \ldots, U_4$ and complex numbers $c_1, \ldots, c_4$ such that

$$A = \sum_{s=1}^{4} c_s U_s$$

(see, e.g., [6, p. 4]).

Then, for $\xi a = 0$, we have again

$$\xi(Aa) = \xi\left(\sum_s c_s U_s a\right) = \sum_s c_s \tilde{U}_s \xi a = 0.$$

Moreover,

$$\|\tilde{A} \xi a\| \leq \sum_{1}^{4} |c_s| \|\xi a\|,$$

so $\tilde{A}$ can be extended to the linear, bounded operator acting in K. It is easy to check that the function η,

$$\eta = A \to \tilde{A}, \qquad (5.29)$$

is a *-representation of $L(H)$ in K.

Let us remark now that

$$\xi P \in \text{range } \tilde{P} \qquad \text{for all} \quad P \in \text{Proj } H \qquad (5.30)$$

and

$$\xi I = \xi P + \xi(I - P). \qquad (5.31)$$

From (5.30) and (5.31) it follows that

$$\xi P = (\eta P)\xi I \qquad \text{for} \quad P \in \text{Proj } H, \qquad (5.32)$$

which ends the proof of sufficiency.

Necessity: Obvious.

5.12 Remarks (a) From the general form of the correlation function of an orthogonally scattered Gleason measure it easily follows that there exist orthogonally scattered Gleason measures which do not fulfill condition (5.26) and, consequently, do not originate from *-representations.

(b) From the proof of Theorem 5.11 it is clear that any orthogonally scattered measure defined on a σ-field of measurable sets originates from a spectral measure. Indeed, let (Ω, S) be a measurable space and $\xi: S \to H$ be an orthogonally scattered measure, $\mu(\cdot) = \|\xi(\cdot)\|^2$. Of course, the formula $(\xi A, \xi B) = \mu(AB) = (\xi(AB), \xi\Omega)$ holds for any $A, B \in S$. ξ can be extended to a linear isomorphism between $L^2_\mu(\Omega, S)$ and H. The spectral measure η can be defined as follows. For $B \in S$ and $f \in L^2_\mu(\Omega, S)$, we put $(\eta B)\xi f = \xi(1_B f)$.

(c) A spectral Gleason measure can be treated as an orthogonally scattered measure (with values in an algebraic Hilbert space) so for spectral Gleason measures a Bochner–Masani duality theorem can be formulated (analogous to Theorem 4.10). It is worth noting here that two spectral Gleason measures satisfying Bochner–Masani duality conditions are simply Wigner's isomorphisms. It follows from the fact that a Gleason spectral measure $\xi: L(H) \to L(K)$ is basic in $L(K)$ if and only if $\dim \xi = 1$.

VI. Convergence of Gleason Measures

6.1 For a separable (real or complex) Hilbert space H, let us denote by $\mathcal{U}(H)$ a real linear space of all real-valued Gleason measures. We introduce in $\mathcal{U}(H)$ a topology given by the following subbasis of neighborhoods:

$$U(\mu, P, \varepsilon) = \{v \in \mathcal{U}(H): |vP - \mu P| < \varepsilon\}, \qquad (6.1)$$

where $\mu \in \mathcal{U}, P \in \text{Proj } H, \varepsilon \in \mathbb{R}^+$. The above topology (of pointwise convergence over $\text{Proj } H$) will be called a weak topology.

Let us denote by $S(H)$ a real linear space of self-adjoint trace-class operators acting in H. $S(H)$ is a Banach space under the norm $\|M\| = \text{tr}|M|$. There is one-one correspondence between $\mathcal{U}$ and S given by the Gleason–Sherstnev formula

$$\mu P = \text{tr } M_\mu P, \qquad P \in \text{Proj } H. \tag{6.2}$$

We shall express weak convergence in $\mathcal{U}(H)$ in terms of corresponding trace-class operators from $S(H)$.

6.2 A family $\{M_t : t \in T\} \subset S(H)$ is said to be τ-compact if the following conditions are satisfied:

(1) $\sup_t \text{tr}|M_t| < \infty$;
(2) for every orthonormal basis $\{e_k\}$ in H, the series $\sum_k |M_t e_k, e_k\rangle|$ is convergent uniformly with respect to t.

6.3 Definition A sequence $\{M_\alpha\}$ of operators from $S(H)$ is said to be τ-convergent to M iff

(a) the set $\{M_\alpha\}$ is compact, and
(b) $M_\alpha \to M$ weakly.

6.4 Theorem [14] *A sequence $\{\mu_\alpha\}$ of Gleason measures from $\mathcal{U}$ is weakly convergent if and only if the sequence of corresponding operators $\{M_{\mu_\alpha}\}$ is τ-convergent.*

Proof Let $\mu_\alpha \to \mu$. Then, in particular, $\mu_\alpha[e] \to \mu[e]$ for $e \in \Delta$ = unit sphere in H, so $(M_{\mu_\alpha} e, e) \to (M_\mu e, e)$, i.e., $M_{\mu_\alpha} \to M_\mu$ weakly. Let $\{c_k\}$ be an arbitrary zero–one sequence and $\{e_n\}$ be an orthonormal basis is H. Let us put $P = \sum_k c_k[e_k]$. Since the limit $\lim_\alpha \sum_k \mu_\alpha[e_k] \cdot c_k$ exists for every zero–one sequence $\{c_k\}$, then by Schur's theorem [10, 37] we have

$$\sup_\alpha \sum_k |\mu_\alpha[e_k]| = \sup_\alpha \sum_k |(M_{\mu_\alpha} e_k, e_k)| < \infty$$

and the series $\sum_k |(M_{\mu_\alpha} e_k, e_k)|$ is convergent uniformly with respect to α.

If the sequence $\{M_{\mu_\alpha}\}$ is τ-convergent to M then, in particular, $M_{\mu_\alpha} \to M$ weakly, which means that $\mu_\alpha[e] \to \mu[e]$ for any $e \in \Delta$. The compactness of $\{M_{\mu_\alpha}\}$ guarantees then the convergence $\mu_\alpha P \to \mu P$ for every $P \in \text{Proj } H$.

6.5 Let us remark that the cone $\mu^+(H)$ of all nonnegative Gleason measures (with the topology at weak convergence) is a metrizable complete space. In fact, a complex Hilbert space H can only be treated over the field of reals (and then it will be denoted by H_1) with an inner product $(x, y)_1 = \text{Re}(x, y)$. To any s-operator M acting in H we can associate a Gaussian measure in H_1 in the following way. We put $f(h) = \exp(-1/2(Mh, h))$ for

$h \in H_1$. Then f is the characteristic functional (Fourier transform) of a Gaussian measure G_M in H_1. The weak convergence of measures in a metric space is metrizable (Lévy–Prokhorov metric [30, 34]). Thus we can define the distance ρ between two s-operators M_1 and M_2 by putting

$$\rho(M_1, M_2) = d(G_{M_1}, G_{M_2}), \tag{6.3}$$

where d is the Lévy–Prokhorov distance.

Let us also remark that Theorem 6.4 can be formulated (and proved) for complex-valued Gleason measures (cf. Section II).

6.6 A family of s-operators $\{M_x : x \in K\}$ is called consistent when the following conditions are satisfied: (α) $\operatorname{tr} M_x = \|x\|^2$; ($\beta$) $M_{x+y} + M_{x-y} = 2M_x + M_y$; ($\gamma$) $M_{\lambda x} = |\lambda|^2 M_x$ for all $x, y \in K$, and $\lambda \in \mathbb{C}$. If ξ is a semispectral Gleason measure, then for every $x \in K$ the mapping $P \to (\xi(P)x, x) = M_x(P)$, $P \in \operatorname{Proj} H$, is an ordinary positive Gleason measure, so there exists an s-operator M_x such that $\mu_x(P) = \operatorname{tr} M_x P$ for $P \in \operatorname{Proj} H$. To a given ξ we can associate the family of Gleason measures $\{\mu_x : x \in K\}$ or the family of corresponding density operators $\{M_x : x \in K\}$. The conditions (α), (β), and (γ) mean that the operators $\{M_x : x \in K\}$ originate from a semispectral Gleason measure ξ such that $(\xi(P)x, x) = \operatorname{tr} M_x P$ for $P \in \operatorname{Proj} H, x \in K$. The same consistency can be written in terms of measures:

(α') $\mu_x(I) = \|x\|^2$;
(β') $\mu_{\lambda x} = |\lambda|^2 \mu_x$;
(γ') $\mu_{x+y} + \mu_{x-y} = 2(\mu_x + \mu_y)$.

Then the formula $\mu_x(\cdot) = (\xi(\cdot)x, x)$ defines ξ uniquely. It can be easily shown that ξ is a spectral measure if and only if

(δ') $\mu_x(\cdot) = \sum_k |\beta_{x, e_k}(\cdot)|^2$ for any $x \in K$ and orthonormal basis $\{e_k\}$, where β is given by the polarization formula

$$\beta_{x, y} = 1/4(\mu_{x+y} - \mu_{x-y} + i\mu_{x+iy} - i\mu_{x-iy}).$$

Let us remark that the conditions (α'), (β'), and (γ') are also sufficient for a system of Gleason measures $\{\mu_x : x \in K\}$ to originate from some semispectral Gleason measure. First we assume K to be a real Hilbert space. Let us put for $P \in \operatorname{Proj} H$

$$B_P(x, y) = \beta_{x, y}(P), \qquad x, y \in K. \tag{6.4}$$

By standard reasoning (see, e.g., [8]) we check that, for a fixed $P \in \operatorname{Proj} H$, B_P is a symmetric, bounded bilinear form. Thus $B_P(x, y) = (\xi P x, y)$, where ξP is a self-adjoint operator such that $0 \leq \xi P \leq I$. Moreover, by (α') we have $\xi I = I$. Since $(\xi P x, x)$ is a Gleason measure for any $x \in K$, $\xi : \operatorname{Proj} H \to L^+(K)$ is a semispectral Gleason measure.

The conditions (α'), (β'), (γ'), and (δ') are sufficient for a system $\{\mu_x : x \in K\}$ to originate from a spectral Gleason measure. In fact, from (δ') we have

$$(\xi Px, x) = \sum_k |(\xi Px, e_k)|^2 = \|\xi Px\|^2 = ((\xi P)^2 x, x). \tag{6.5}$$

6.7 For two separable Hilbert spaces H and K, let us denote by $\pi(H, K)$ the space of all semispectral Gleason measures $\xi : L(H) \to L^+(K)$. The following two topologies in $\pi(H, K)$ seem to be natural.

(α) A weak topology: given by the subbasis of neighborhoods

$$\mathscr{U}(\eta, P, x, \varepsilon) = \{\xi \in \pi : |(\xi Px, x) - (\eta Px, x)| < \varepsilon\},$$

where $\eta \in \pi(H, K)$, $P \in \operatorname{Proj} H$, $x \in L$, $\varepsilon \in \mathbb{R}^+$.

(β) A strong topology: given by the subbasis of neighborhoods

$$\mathscr{V}(\eta, P, x, \varepsilon) = \{\xi \in \pi : \|\xi Px - \eta Px\|_K < \varepsilon\},$$

where $\eta \in \pi$, $P \in \operatorname{Proj} H$, $x \in K$, $\varepsilon \in \mathbb{R}^+$.

In other words, a sequence $\{\xi_n\}$ of semispectral Gleason measures is weakly (strongly) convergent to a semispectral measure ξ if for any projector $P \in \operatorname{Proj} H$ we have $\xi_n(P) \to \xi(P)$ in the weak (strong) operator topology in $L(K)$.

6.8 Theorem *The following conditions are equivalent ($\rightharpoonup$ denotes the convergence in the weak operator topology).*
  (1) $\xi_n \to \xi$ weakly.
  (2) $\xi_n(a) \rightharpoonup \xi(a)$ for all $a \in L(H)$.
  (3) $\xi_n(u) \rightharpoonup \xi(u)$ for all unitary operators.
  (4) For any $x \in K$ and any resolution of the identity $E(\cdot)$ in H, the sequence of measures $(\xi_n(E(\cdot))x, x)$ converges weakly to the measure $(\xi(E(\cdot))x, x)$.
  (5) $\mu_x^n(P) \to \mu_x(P)$ for $x \in K$, $P \in \operatorname{Proj} H$, where $\mu_x^n(\cdot) = (\xi_n(\cdot)x, x)$.
  (6) $\mu_x^n[e] \to \mu_x[e]$, $x \in K$, $e \in \Delta = $ unit sphere in H, and the series $\sum_k \mu_k^n(\hat{e}_k)$ is uniformly convergent with respect to n.
  (7) Putting $(\xi_n(P)x, x) = \operatorname{tr} M_n^x P$, the sequence of s-operators $\{M_n^x\}_{n=1}^\infty$ is τ-convergent to an s-operator M^x for every $x \in K$ and the family $\{M_x : x \in K\}$ is consistent.
  (8) For $x \in K$, $\rho(M_n^x, M^x) \to 0$ as $n \to \infty$.
  (9) $\delta(\xi_n, \xi) \to 0$,
where $\delta(\xi, \eta) = \sum_k 2^{-k} \rho(M_\xi^{x_k}, M_\eta^{x_k})$, $\{x_k\}$-dense in K.

We can omit the detailed proof (see [18]).

From condition (9) of Theorem 6.8 it follows, in particular, that the weak convergence of semispectral measures is metrizable. It is worth noting that for spectral Gleason measures the strong convergence is, in some sense, most natural. Namely, we have the following theorem.

6.9 Theorem *If a sequence $\{\xi_n\}$ of spectral Gleason measures converges weakly to a spectral measure ξ, then ξ_n converges to ξ strongly*

Proof Let $\xi_n : \operatorname{Proj} H \to \operatorname{Proj} K$ ($n = 1, 2, \ldots$) be a sequence of spectral Gleason measures and let ξ_n converge weakly to a spectral measure ξ. Then, by Theorem 6.8 (condition (3)) we have $\xi_n U \to \xi U$ (in a weak operator topology) for all unitary operators acting in H. Let us notice now that any spectral Gleason measure transforms unitary operators into unitary operators. Namely, if $U = \int e^{i\lambda} E(d\lambda)$ is a spectral decomposition of a unitary operator U and η is a spectral Gleason measure, then $\eta U = \int e^{i\lambda}(\eta E)(d\lambda)$ is a spectral decomposition of a unitary operator ηU. Evidently, if a sequence V_n of unitary operators converges weakly to a unitary operator V, then $V_n \to V$ strongly. In our case it gives us $\xi_n U \to \xi U$ strongly for any unitary operator U. Since every operator $A \in L(H)$ can be represented as a linear combination of four unitary operators [6], it follows that $\xi_n \to \xi$ strongly. This concludes the proof of the theorem.

6.10 For a Hilbert space H and a Banach space X, let us denote by $\mathscr{U}(H, X)$ a linear space of all X-valued Gleason measures defined on $\operatorname{Proj} H$. We introduce in $\mathscr{U}(H, X)$ a Banach space topology putting for $\xi \in \mathscr{U}$:

$$\|\xi\| = \sup\{\|\xi P\| : P \in \operatorname{Proj} H\}. \tag{6.6}$$

Let $M_\xi : X^* \to C_1(H)$ be a linear operator associated with $\xi \in \mathscr{U}(H, X)$ by Theorem 2.6; thus we have, in particular,

$$\langle \xi P, x^* \rangle = \operatorname{tr} M(x^*)P, \quad x^* \in X^*, \quad P \in \operatorname{Proj} H. \tag{6.7}$$

Let us remark now that

$$\|\xi\| \leq \||M_\xi|\| \leq 4\|\xi\|, \tag{6.8}$$

where $\||M_\xi|\|$ denotes a norm of M_ξ in $L(X^*, C_1(H))$, where X^* and $C_1(H)$ are endowed with the norm topologies (see 2.5 and 2.6). In fact, by (2.17) we have

$$\|\xi P\| = \sup_{P \in \operatorname{Proj} H} \sup_{\|x^*\| = 1} |\langle \xi P, x^* \rangle|$$

$$= \sup_{\|x^*\| = 1} \sup_{P \in \operatorname{Proj} H} |\operatorname{tr} M(x^*)P| \leq \||M_\xi|\|, \tag{6.9}$$

and by (2.16)

$$\||M_\xi|\| = \sup_{\|x^*\| = 1} \|M(x^*)\| \leq 4 \sup_{\|x^*\| = 1} \sup_{P \in \operatorname{Proj} H} |\langle \xi P, x^* \rangle| \leq 4\|\xi\|. \tag{6.10}$$

Hence, we obtain

6.11 Theorem *The following conditions are equivalent for* $\{\xi_n\} \subset \mathcal{U}(H, X)$.

(i) $\|\xi_n\| \to 0$.
(ii) $\||M_{\xi_n}\|| \to 0$.
(iii) $\sup\{|\langle \xi P, x^* \rangle| : P \in \operatorname{Proj} H, \|x^*\| = 1\} \to 0$.

6.12 Corollary *A sequence* $\mu_n P = \operatorname{tr} M_n P$ *of real-valued Gleason measures is uniformly (on* $\operatorname{Proj} H$*) convergent to zero if and only if*

$$\lim_n \operatorname{tr}|M_n| = 0. \tag{6.11}$$

6.13 From the proof of Theorem 6.11 and Theorem 2.6, it follows that, if a Banach space X is reflexive, then the Banach spaces $\mathcal{U}(H, X)$ and $L(X^*, C_1(H))$ are ismorphically homeomorphic. The isomorphism is given by $\xi \to M_\xi$.

VII. Gleason Measures in Tensor Products

7.1 Let $\{H_j\}$ be a sequence of copies of a separable Hilbert space H and let us fix a sequence $e = \{e_j\}$ of unit vectors in H. Let us denote by $\tilde{H} = \bigotimes_1^\infty{}^e H_j$ the von Neumann (incomplete) tensor product of H_j [27].

Let μ_n be a positive real-valued Gleason measure defined on the tensor product $\tilde{H}_n = \bigoplus_1^n H_i$. The sequence $\{\mu_n\}$ is said to be consistent if for any positive integer n and any $P_n \in \operatorname{Proj} \tilde{H}_n$ the formula

$$\mu_{n+1}(P_n \otimes I) = \mu_n(P_n) \tag{7.1}$$

holds (I denotes the identity operator in H). In general there is no extension of a "cylindrical" Gleason measure in $\bigotimes_1^\infty{}^e H_i$ represented by $\{\mu_n\}$ to a Gleason measure defined on $\operatorname{Proj} \tilde{H}$ (cf. [28]), i.e., there does not exist a measure μ on $\operatorname{Proj} \tilde{H}$ such that

$$\mu\left(P_n \otimes \bigotimes_{n+1}^\infty I\right) = \mu_n(P_n) \quad \text{for all } n \text{ and } P_n \in \operatorname{Proj} \tilde{H}_n. \tag{7.2}$$

We now formulate the necessary and sufficient condition for the existence of this extension μ.

Let us denote by S_n the density operator corresponding to μ_n, i.e., $\mu_n P_n = \operatorname{tr} S_n P_n$ for $P_n \in \operatorname{Proj} \tilde{H}_n$. Let us define a sequence of s-operators in $\tilde{H}$, putting

$$Q_n = S_n \otimes \bigotimes_{i=n+1}^{\infty}{}^e [e_i]. \tag{7.3}$$

7.2 Theorem *A cylindrical Gleason measure in $\tilde{H} = \bigotimes_{1}^{\infty}{}^e H_j$ given by a consistent sequence of measures $\{\mu_n\}$ can be extended to a Gleason measure μ defined on* $\operatorname{Proj} \tilde{H}$ *if and only if the sequence $\{Q_n\}$ given by (7.3) is τ-convergent (automatically to an s-operator).*

Proof Sufficiency: Let $Q_n \xrightarrow{\tau} S$. Then the formula

$$\mu P = \operatorname{tr} SP, \qquad P \in \operatorname{Proj} \tilde{H}, \tag{7.4}$$

defines a Gleason measure in $\tilde{H}$. Let us put

$$v_n P = \operatorname{tr} Q_n P \quad \text{for} \quad P \in \tilde{H}. \tag{7.5}$$

Then, by Theorem 6.4, $v_n \to \mu$ weakly. But for $n > k$, by (7.1) we have for $P = P_k \otimes \bigotimes_{k+1}^{\infty} I$ and $P_k \in \operatorname{Proj} \tilde{H}_k$

$$v_n P = \mu_n \left(P_k \otimes \bigotimes_{k+1}^{n} I \right) \cdot \operatorname{tr} \left(\bigotimes_{n+1}^{\infty}{}^e [e_i] \cdot \bigotimes_{n+1}^{\infty}{}^e I \right)$$

$$= \mu_k P_k \cdot \operatorname{tr} \left(\bigotimes_{n+1}^{\infty} [e_i] \cdot \bigotimes_{n+1}^{\infty}{}^e I \right). \tag{7.6}$$

A simple computation shows that

$$\operatorname{tr} \left(\bigotimes_{n+1}^{\infty}{}^e [e_i] \cdot \bigotimes_{n+1}^{\infty}{}^e I \right) \to 1, \tag{7.7}$$

so $v_n P \to \mu_k P_k$. On the other hand, $v_n P \to \mu P$, so we obtain (7.2). This concludes the proof of sufficiency.

Necessity: Let μ be a measure on $\operatorname{Proj} \tilde{H}$ satisfying condition (7.2). Then by (7.6) and (7.7) for $P = P_k \otimes \bigotimes_{k+1}^{\infty} I$ and $P_k \in \operatorname{Proj} \tilde{H}_n$, we have

$$v_n P \to \mu_k P_k = \mu \left(P_k \otimes \bigotimes_{k+1}^{\infty} I \right) \quad \text{as} \quad n \to \infty. \tag{7.8}$$

This implies that $v_n P \to \mu P$ for all $P \in \operatorname{Proj} \tilde{H}$, i.e., $v_n \to \mu$ weakly. By Theorem 6.4 it entails the τ-convergence of Q_n, which ends the proof.

Let us remark that the sequence of operators Q_n defined by (7.3) is always weakly convergent in $\tilde{H}$.

7.3 Theorem [1] *A consistent sequence $\{\mu_n\}$ of Gleason measures can be extended to a Gleason measure μ on* $\operatorname{Proj} \tilde{H}$ *if and only if*

$$\operatorname{tr} S = \operatorname{tr} S_1, \tag{7.9}$$

where S denotes a weak limit of Q_n.

Proof A standard computation shows that

$$\operatorname{tr} S_n = \operatorname{tr} Q_n = \text{const} \qquad (n = 1, 2, \ldots). \tag{7.10}$$

Then weak convergence $Q_n \to S$, (7.9), and (7.10) give τ-convergence of Q_n. It follows from the fact that if $a_{nk} \geq 0$, $a_{nk} \to a_k$, $\sum_{k=1}^{\infty} a_{nk} \equiv \sum_{k=1}^{\infty} a_k$, then the series $\sum_k a_{nk}$ is convergent uniformly with respect to n. Of course, τ-convergence of Q_n to S implies (7.9). Now it suffices to apply Theorem 7.2.

7.4 Let $\{\mu_i\}$ be a sequence of probability Gleason measures on $\operatorname{Proj} H$. We say that a measure μ on $\operatorname{Proj} \tilde{H}$ is a product of μ_j (in symbols $\mu = \prod \mu_j$) if for any system of projectors $P_1, P_2, \ldots, P_n \in \operatorname{Proj} H$ we have

$$\mu\left(P_1 \otimes P_2 \otimes \cdots \otimes P_n \otimes \bigotimes_{n+1}^{\infty} I\right) = \mu_1 P_1 \mu_2 P_2 \cdots \mu_n P_n. \tag{7.11}$$

Of course, for a finite system $\mu_1, \mu_2, \ldots, \mu_n$ of Gleason measures we have $\mu = \prod_1^n \mu_j$ iff $S_\mu = \bigotimes_1^n S_{\mu_j}$, where S_α denotes the density operator of a measure μ_α. This remark and Theorem 7.3 give us the following result.

7.5 Theorem [1, 5] *For a sequence of probability Gleason measures μ_j on* $\operatorname{Proj} H$, *where $\mu_j P = \operatorname{tr} S_j P$, $P \in \operatorname{Proj} H$, the infinite product $\prod_j \mu_j$ in $\tilde{H}$ exists if and only if the condition*

$$\prod_{i > N} (S_i e_i, e_i) > 0 \qquad \text{for some} \quad N \tag{7.12}$$

is satisfied.

7.6 The extension theorems for real-valued positive Gleason measures offer a possibility of formulating the extension theorems for the positive operator-valued and orthogonally scattered measures. We leave to the reader the formulation of these results.

7.7 A positive Gleason measure ν on $\operatorname{Proj} \tilde{H}$ is said to be symmetric when

$$\nu\left(P_1 \otimes P_2 \otimes \cdots \otimes P_n \otimes \bigotimes_{n+1}^{\infty} I\right) = \nu\left(P_{i_1} \otimes P_{i_2} \otimes \cdots \otimes P_{i_n} \otimes \bigotimes_{n+1}^{\infty} I\right) \tag{7.13}$$

for any positive integer n, $P_1, \ldots, P_n \in \operatorname{Proj} H$, and an arbitrary permutation $\{i_1, \ldots, i_n\}$ of $\{1, 2, \ldots, n\}$. Of course, every product measure $\tilde{\mu}$

$$\tilde{\mu} = \prod_1^{\infty} \mu_j, \tag{7.14}$$

where μ_j is a sequence of copies of a probability Gleason measure, is a symmetric Gleason measure.

For symmetric Gleason measures we can formulate the following theorem, which is an extension of the classical result of Hewitt and Savage [13] on measures in Cartesian products.

7.8 Theorem [17] *For every symmetric Gleason measure v in $\tilde{H}$ there exists a finite measure q on the space $\tilde{\mathcal{M}}$ of all probability product measures in $\tilde{H}$ such that the formula*

$$vP = \int_{\tilde{\mathcal{M}}} \tilde{\mu}P q(d\tilde{\mu}) \qquad (7.15)$$

holds for every $P \in \operatorname{Proj} \tilde{H}$.

That is, roughly speaking, every symmetric Gleason measure is a mixture of probability product measures. We omit the proof based on the Choquet's theorem [4, 33]. The details can be found in Jajte [17].

VIII. Random Gleason Measures

8.1 Let $(\Omega, \mathcal{T}, m)$ be a probability space and let $L_0(\Omega, \mathcal{T}, m)$ denote a space of real random variables on $(\Omega, \mathcal{T}, m)$ with the topology of convergence in probability m.

By a random independently scattered Gleason measure we mean a stochastic process $\xi: \operatorname{Proj} H \to L_0(\Omega, \mathcal{T}, m)$ such that for every sequence $P_1, P_2, \ldots$ of pairwise orthogonal projectors the random variables $\xi P_1, \xi P_2, \ldots$ are stochastically independent and the equality

$$\xi\left(\sum_j P_j\right) = \sum_j \xi P_j \quad \text{(in law)} \qquad (8.1)$$

holds.[1] In the sequel we shall always assume that all the random variables ξP, $P \in \operatorname{Proj} H$, are symmetric and infinitely divisible. Random Gleason measures may be treated as a noncommutative version of stochastic processes with independent increments.

Let us denote by $\hat{\xi}(P, \cdot)$ the characteristic function (Fourier transform) of ξP, i.e.,

$$\hat{\xi}(P, t) = \mathbb{E}(\exp it\xi P), \qquad t \in \mathbb{R}. \qquad (8.2)$$

8.2 Examples (a) Let μ be a nonnegative finite Gleason measure on $\operatorname{Proj} H$ and let

$$\hat{\xi}(P, t) = \exp(-\mu P \cdot |t|^r), \quad r \leq 2. \qquad (8.3)$$

---

[1] $\zeta = \eta$ (in law) means that the probability distributions of ζ and η are equal to each other.

It follows from Kolmogorov's consistency theorem that given a "control measure" μ one can construct the related independently scattered random Gleason measure with the characteristic functional (8.3) (stable symmetric Gleason measure of exponent r). For the Gaussian case we have the following formula for a correlation function of our stochastic process:

$$\mathbb{E}\xi P \xi Q = \operatorname{tr} MPQ, \tag{8.4}$$

where M is the density operator of the "control measure" μ, i.e., $\mu(\cdot) = \operatorname{tr} M(\cdot)$. It follows immediately from Theorem 3.4 on orthogonally scattered Gleason measure.

By Theorem 5.4 and 5.5 for a symmetric Gaussian Gleason measure ξ we have a representation

$$\xi P = (\eta P) X_0, \tag{8.5}$$

where η is a spectral Gleason measure (extendable to a *-representation of $L(H)$ in $L_2(\Omega, \mathcal{T}, m)$), and $X_0 = \xi(I)$.

(b) For $\hat{\xi}(P, t) = \exp[\mu P \cos(t - 1)]$ there exists a (Poissonian) Gleason measure ξ with the characteristic function $\hat{\xi}$ as above (given by the Gleason measure μ).

8.3 We give now a Lévy–Khinchine spectral representation for random Gleason measures. Theorems 8.4 and 8.6 give the description of a random Gleason measure in terms of s-operators families. In particular, an operator-valued measure N (Theorem 8.6) plays the same role as the Lévy–Khinchine spectral measure in the theory of infinitely divisible distributions and it is a very convenient tool in considerations concerning convergence and compactness of sequences of random Gleason measures.

8.4 Theorem [12] *Let ξ be a random Gleason measure. Then there exists a one-parameter family M_t, $t \in \mathbb{R}$, of s-operators such that*

$$\hat{\xi}(P, t) = \exp(-\operatorname{tr} M_t P), \qquad P \in \operatorname{Proj} H, \tag{8.6}$$

and, moreover, the family M_t satisfies the following conditions:

(α) $M_0 = 0$, $M_{-t} = M_t$, $t \in \mathbb{R}$;
(β) $\{M_t : t \in \mathbb{R}\}$ *is locally τ-compact, i.e., for every $T > 0$, the family $\{M_t : |t| < T\}$ is τ-compact (in the sense of Definition 6.2);*
(γ) $\lim_{t \to 0} (M_t x, x) = 0$ *for $x \in H$;*
(δ) *the quadratic form $(M_t x, x)$ is negative definite, i.e., for each sequence $\alpha_1, \ldots, \alpha_n$ of complex numbers such that $\sum_1^n \alpha_j = 0$, we have*

$$\sum_{j,k=1}^{n} (M_{t_j - t_k} x, x) \alpha_j \bar{\alpha}_k \leq 0, \qquad t_j \in \mathbb{R}, \quad x \in H.$$

Conversely, if the family $\{M_t : t \in \mathbb{R}\}$ *of s-operators satisfies the conditions* (α)–(δ), *then there exists a random Gleason measure* ξ *such that* (8.6) *holds.*

Proof Let ξ be a Gleason measure and let us consider a sequence of pairwise orthogonal projectors $P_1, P_2, \ldots$. Denoting $P = \sum_1^\infty P_k$ and using the independence of ξP_k, we obtain

$$\hat{\xi}(P, t) = \lim_{n \to \infty} \prod_{k=1}^n \hat{\xi}(P_k, t) \tag{8.7}$$

for all $t \in \mathbb{R}$. Since ξP is infinitely divisible, we have $\hat{\xi}(P, t) > 0$ (see, e.g., [22]). Putting

$$\psi(P, t) = -\ln \hat{\xi}(P, t) \quad \text{for} \quad P \in \operatorname{Proj} H \tag{8.8}$$

we have $\psi \geq 0$ and

$$\psi(P, t) = \sum_1^\infty \psi(P_k, t). \tag{8.9}$$

Therefore, $\psi(\cdot, t)$ is a Gleason measure for any $t \in \mathbb{R}$ and, by the Gleason theorem, we can write

$$\psi(P, t) = \operatorname{tr} M_t P, \quad t \in \mathbb{R}, \quad P \in \operatorname{Proj} H, \tag{8.10}$$

where M_t is an s-operator in H. It remains to show (α)–(δ). (α) is obvious by the symmetry of ξP. To show (β), let us notice that for a given $T > 0$ there exists $\delta > 0$ such that $\inf_{|t| \leq T} \hat{\xi}(I, t) \geq \delta > 0$. Thus we have $\sup_{|t| \leq T} \operatorname{tr} M_t \leq -\ln \delta < \infty$. Let $\{e_k\}$ be an orthonormal basis in H. Then putting $P_n = \sum_{n+1}^\infty [e_k]$, we obtain

$$\operatorname{tr} M_t P_n = \sum_{k=n+1}^\infty (M_t e_k, e_k). \tag{8.11}$$

On the other hand, $\xi P_n \to 0$ in probability. Hence

$$\lim_{n \to \infty} \inf_{|t| \leq T} \hat{\xi}(P_n, t) = 1 \tag{8.12}$$

and

$$\lim_{n \to \infty} \sup_{|t| \leq 1} \operatorname{tr} M_t P_n = 0. \tag{8.13}$$

The condition (γ) follows from the continuity of $\hat{\xi}(P, \cdot)$. The quadratic form $(M_t x, x)$ is negative definite by Johansen [20]. This concludes the proof of (α)–(δ).

Conversely, suppose that a family $\{M_t : t \in \mathbb{R}\}$ of s-operators satisfies the conditions (α)–(δ). Let us put

$$\varphi(P, t) = \exp(-\operatorname{tr} M_t P), \quad P \in \operatorname{Proj} H. \tag{8.14}$$

By Johansen [20] and the assumptions on $\{M_t\}$, $\varphi(P, \cdot)$ is a characteristic function of a symmetric infinitely divisible distribution. For a finite sequence of projectors $P_1, P_2, \ldots, P_n$, let us put

$$\hat{q}_{P_1,\ldots,P_n}(t_1,\ldots,t_k) = \prod_{j=1}^{n} \varphi_{P_j}(t_j) = \exp(-\operatorname{tr} M_{t_j} P_j) \qquad (8.15)$$

as the characteristic function of a probability distribution $q_{P_1,\ldots,P_n}$ in $\mathbb{R}^k$. By Kolmogorov's consistency theorem, there exists a stochastic process $\{\xi P : P \in \operatorname{Proj} H\}$ over some probability space $(\Omega, \mathcal{T}, m)$ such that the joint distribution of a random vector $\{\xi P_1, \ldots, \xi P_n\}$ is $q_{P_1,\ldots,P_n}$. In particular, we have (8.6). Let $P_1, P_2, \ldots$ be a sequence of pairwise orthogonal projectors. From the form of the characteristic function we conclude that $\xi P_1, \xi P_2, \ldots$ are independent and we have $\xi P_1 + \cdots + \xi P_n = \xi(\sum_1^n P_k)$ (in law). Putting $P = \sum_1^\infty P_k$ we obtain for every $t \in \mathbb{R}$

$$\hat{\xi}\left(\sum_1^n P_k, t\right) = \exp\left(-\operatorname{tr} M_t \sum_1^n P_k\right) \to \exp(-\operatorname{tr} M_t P) = \hat{\xi}(P, t).$$

Hence, $\sum_1^n \xi P_k \to \xi P$ in $L_0(\Omega, \mathcal{T}, m)$. Thus ξ is a random Gleason measure satisfying (8.6). The proof of the theorem is completed.

8.5 Let N be a function defined on the Borel σ-field B in $\mathbb{R}^+$ taking its values in the space of s-operators. N is said to be an s-measure if for any sequence $B_1, B_2, \ldots$ of pairwise disjoint Borel sets in $\mathbb{R}^+$, the sequence $\sum_{j=1}^{n} N(B_j)$ is τ-convergent to $N(\bigcup_{j=1}^{\infty} B_j)$.

8.6 Theorem [12] *Let ξ be a random Gleason measure. Then*

$$\hat{\xi}(P, t) = \exp\left\{-\int_0^\infty (1 - \cos \lambda t)[(1 + \lambda^2)/\lambda^2] \operatorname{tr} N(d\lambda) P\right\}, \qquad (8.16)$$

where N is an s-measure (in the sense of 8.5) defined on the Borel field $\mathcal{B}(\mathbb{R}^+)$. Conversely, if N is an s-measure, then there exists a random Gleason measure ξ such that (8.16) holds.

Proof Let ξ be a random Gleason measure. By the Lévy–Khinchine formula we can write

$$\hat{\xi}(P, t) = \exp\left\{-\int_0^\infty (1 - \cos \lambda t) \mu(P, d\lambda)\right\}, \qquad (8.17)$$

where $\mu(P, \cdot)$ is a positive finite Borel measure on $\mathcal{B}(\mathbb{R}^+)$. For any fixed $B \in \mathcal{B}(\mathbb{R}^+)$, $\mu(\cdot, B)$ is a Gleason measure on $\operatorname{Proj} H$. Indeed, for a sequence $P_1, P_2, \ldots$ of pairwise orthogonal projectors we have, by the uniqueness

GLEASON MEASURES

of the Lévy–Khinchine representation,

$$\sum_{k=1}^{n} \mu(P_k, \cdot) = \mu\left(\sum_{k=1}^{n} P_k, \cdot\right). \tag{8.18}$$

Moreover, (8.1) implies the weak convergence of measures $\mu(\sum_1^\infty P_k, \cdot)$ to the measure $\mu(\sum_1^\infty P_k, \cdot)$. In particular, $\mu(\sum_1^n P_k, \mathbb{R}^+) \to \mu(\sum_1^\infty P_k, \mathbb{R}^+)$, so the series $\sum_1^\infty \mu(P_k, B)$ is convergent uniformly with respect to $B \in \mathscr{B}(\mathbb{R}^+)$. We can define the measure v, putting $v(B) = \sum_1^\infty \mu(P_k, B)$ for $B \in \mathscr{B}(\mathbb{R}^+)$. By the uniqueness of the weak limit we have $v = \mu(\sum_1^\infty P_k, \cdot)$, so $\sum_1^\infty \mu(P_k, B) = \mu(\sum_1^\infty P_k, B)$ holds for all $B \in \mathscr{B}(\mathbb{R}_+)$. The Gleason theorem gives

$$\mu(P, B) = \operatorname{tr} N(B)P \tag{8.19}$$

for some s-operator $N(B)$. One can show that N is an s-measure. We omit the details, which can be found in Hensz and Jajte [12].

If now N is an s-measure on $\mathbb{R}_+$, then for a fixed $P \in \operatorname{Proj} H$ the formula

$$\mu(P, t) = \int_0^\infty (1 - \cos \lambda t)[(1 + \lambda^2)/\lambda^2] \operatorname{tr} N(d\lambda)P \tag{8.20}$$

describes the logarithm of a symmetric infinitely divisible characteristic function. From that fact it easily follows that for any fixed $t \in \mathbb{R}_+$, $P \to \mu(P, t)$ is a Gleason measure. Let us put

$$\mu(P, t) = \operatorname{tr} M_t P, \qquad P \in \operatorname{Proj} H. \tag{8.21}$$

One can verify that the family $\{M_t : t \in \mathbb{R}_+\}$ satisfies the conditions (α)–(δ) of the Theorem 8.4, which ends the proof.

8.7 We shall examine now some relationships between random fields over $L(H)$ and Gleason measures. The following definition is suggested by the theory of stationary processes (cf. [3]). Let H, K be Hilbert spaces [we obtain a "probabilistic version," putting $K = L_2(\Omega, \mathscr{T}, m)$].

8.8 Definition A mapping $X : L(H) \to K$ is said to be a stationary field of the second order over the algebra $L(H)$ when the formula

$$(X(a), X(b)) = \varphi(b^*a) \tag{8.22}$$

holds for every $a, b \in L(H)$ and some functional φ on $L(H)$. The functional φ is called the correlation functional of X.

8.9 Theorem *A stationary field X over $L(H)$ is an orthogonally scattered Gleason measure if and only if its correlation functional φ is linear and continuous in the strong operator topology in $L(H)$. Then there exist a *-quasi-representation η of $L(H)$ in K of $L(H)$ in K and an s-operator M such that*

$$X(a) = \eta(a) \cdot X(I) \qquad \text{and} \qquad \varphi(a) = \operatorname{tr} Ma \qquad \text{for all} \quad a \in L(H). \tag{8.23}$$

Proof Sufficiency: If (8.22) holds, then for any unitary operator U we have

$$(X(Ua), X(Ub)) = \varphi(b^*a) = (X(a), X(b)). \tag{8.24}$$

Then, putting $\tilde{U}X(a) = X(Ua)$, we can define (by a standard extension) a linear (unitary) operator $\tilde{U}$ on the space K (to avoid trivial complications, we assume that $\sigma\{X(a): a \in L(H)\} = K$).

For an arbitrary $A \in L(H)$ we can write A as a linear combination of unitary operators $A = \sum_k c_k U_k$ and then put

$$\tilde{A}X(a) = \sum_k c_k \tilde{U}_k X(a). \tag{8.25}$$

By the linearity of φ, we have

$$X\left(\sum_k c_k U_k a\right) = \sum_k c_k X(U_k a), \tag{8.26}$$

so we can rewrite (8.25) as

$$\tilde{A}X(a) = X(Aa), \quad A \in L(H). \tag{8.27}$$

Then (8.26) can be written as

$$\left(\sum_k c_k U_k\right)^{\sim} = \sum_k c_k \tilde{U}_k. \tag{8.28}$$

Formula (8.27) gives us (after a natural extension) the definition of $\tilde{A}$. By (8.25), $\tilde{A}$ is a linear bounded operator in K.

Now we can define the operation $\eta: L(H) \to L(K)$ by putting

$$\eta: A \to \tilde{A} \quad \text{for} \quad A \in L(H). \tag{8.29}$$

η is a spectral Gleason measure. In fact, we have for $P \in \text{Proj } H$

$$\tilde{P} = \tilde{P}, \quad \tilde{P}^2 = \tilde{P}, \quad \text{and} \quad \tilde{P}Q = \tilde{P}\tilde{Q}.$$

Moreover, by the linearity of φ, we have

$$\eta\left(\sum_k z_k A_k\right) = \sum_k z_k \eta(A_k) \quad \text{for} \quad z_k \in \varphi, \ A_k \in L(H). \tag{8.30}$$

It remains to show that for a sequence $P_1, P_2, \ldots$ of mutually orthogonal projectors from $\text{Proj } H$ we have

$$\eta\left(\sum_{k=1}^{\infty} P_k\right) = \sum_{k=1}^{\infty} \eta(P_k).$$

The last formula follows easily from the additivity of η and from the continuity of φ in the strong operator topology.

From (8.27) and (8.29) we obtain

$$X(a) = \eta(a)X(I), \qquad (8.31)$$

where η, as a spectral Gleason measure, is, by Theorem 5.7, a *-quasi-representation of $L(H)$ in K. The second part of (8.23) is a consequence of the Gleason theorem.

Necessity: Trivial.

ACKNOWLEDGMENTS

This paper was written while the author held a visiting position at Wayne State University, whose hospitality he gratefully acknowledges. The author wishes to thank Professor P. Masani for reading a part of the manuscript of this paper and for suggestions of some improvements.

REFERENCES

1. Bartoszewicz, A., Kolmogorov consistency theorem for Gleason measures, *Colloq. Math.* **39** (1978), 141–151.
2. Bochner, S., Inversion formulae and unitary transformations, *Ann. of Math.* **35** (1934), 111–115.
3. Bochner, S., "Harmonic Analysis and the Theory of Probability." Univ. of California Press, Berkeley, 1955.
4. Choquet, G., Le theorem de representation integral dans les ensembles convexes compacts, *Ann. Inst. Fourier* (Grenoble) **10** (1960), 333–344.
5. Christensen, M., Gleason measures on infinite tensor products of Hilbert spaces, *J. Mathematical Phys.* **18** (1977), 113–115.
6. Dixmier, J., "Les Algèbres d'Opérateurs dans l'Espace Hilbertien (Algebres de von Neumann)." Gauthier-Villars, Paris, 1969.
7. Dixmier, J., "Les C*-Algebres et Leurs Representations." Gauthier-Villars, Paris, 1969
8. Dunford, N., and Schwarz, J., "Linear Operators," Part I. New York, 1958.
9. Gleason, A. M., Measures on the closed subspaces of a Hilbert space, *J. Math. Mech.* **6** (1957), 885–893.
10. Hardy, G. H., "Divergent Series." Oxford, 1949.
11. Hensz, E., Remarks on vector Gleason measures, preprint.
12. Hensz, E., and Jajte, R., Random measures of Gleason type, *Eur. Meet. Statist., 7th Prague Confer, Information Theory, Statist. Decision Functions Random Processes, 1974,* 231–237.
13. Hewitt, E., and Savage, L. J., Symmetric measures on Cartesian products, *Trans. Amer. Math. Soc.* **8** (1956), 470–501.
14. Jajte, R., On convergence of Gleason measures, *Bull. Acad. Pol. Sci.* **20** (1972), 211–214.
15. Jajte, R., and Loebl, R., Gleason measures and some inner production L(H), *Studia Math.* (to appear).
16. Jajte, R., Orthogonally scattered Gleason measures, *Bull. Acad. Pol. Sci.* **26** (1978), 685–689.
17. Jajte, R., Symmetric Gleason measures (to appear).
18. Jajte, R., Spectral and semi-spectral Gleason measures, preprint.
19. Jajte, R., and Paszkiewicz, A., Vector measures on the closed subspaces of a Hilbert space, *Studia Math.* **58** (1978), 229–251.
20. Johansen, S., An application of extreme point methods to the representation of infinitely divisible distributions, *Z. Wahrscheinlichkeitstheorie und Verw. Gebiete* **5** (1966), 304–316.

21. Lodkin, A. A., Every measure on the projectors of a W*-algebra can be extended to a state, *Funkcional Anal. i Priložen* **8** (1974), 54–58 (in Russian).
22. Loève, M., "Probability Theory," 2nd Ed. Wiley, New York, 1957.
23. Mackey, G. W., "The Mathematical Foundations of Quantum Mechanics." Amsterdam, 1963.
24. Masani, P., Orthogonally scattered measures, *Advances in Math.* **2** (1968), 61–117.
24a. Masani, P., Quasi-isometric measures and their applications, *Bull. Amer. Math. Soc.* **76** (1970), 427–528.
25. Naimark, N., "Normed Rings." P. Noordhoff, New York, 1959.
26. von Neumann, J., "Mathematische Grundlagen der Quantenmechanik." Berlin, 1932.
27. von Neumann, J., On infinite direct products, *Compositio Math.* **6** (1938), 1–77.
28. Nowak, B., On a problem in Gleason's measure theory, *Bull. Acad. Pol. Sci.* **22** (1974), 393–395.
29. Orlicz, W., Über unbedingte Konvergenz in Funktionalräumen I, *Studia Math.* **4** (1933), 33–37.
30. Parthasarathy, K. R., "Probability Measures on Metric Spaces." Academic Press, New York, 1967.
31. Paszkiewicz, A., On homomorphisms of projective lattices in complex Hilbert space, *Colloq. Math.* (to appear).
32. Pedersen, G. K., Measure theory for C*-algebras, *Math. Scand.* **19** (1966), 131–145.
33. Phelps, R. R., "Lectures on Choquet's Theorem." Princeton, New Jersey, 1966.
34. Prokhorov, J., Convergence of random processes and limit theorems, *Teor. Verojatnost. i Primenen.* **1** (1956), 177–238.
35. Ringrose, J. R., "Compact Non-Self-Adjoint Operators." Van Nostrand Reinhold, Princeton, New Jersey, 1971.
36. Sakai, S., A Radon-Nikodym theorem in w*-algebras, *Bull. Amer. Math. Soc.* **71** (1965), 149–151.
37. Schur, I., Über lineare Transformationen in der Theorie der unendlichen Reichen, *J. Reine Ange. Math.* **151** (1921), 79–111.
38. Segal, I. E., A non-commutative extension of abstract integration, *Ann. of Math.* **57** (1953), 401–457.
39. Sherstnev, A. N., On signed measures in non-commutative measure theory, Verojatn. Met., *Izd. Kazan. Univ.* **10** (1974), 68–72 (in Russian).
40. Stinespring, W. F., Positive functions on C*-algebras, *Proc. Amer. Math. Soc.* **6** (1955), 211–216.
41. Størmer, E., Positive linear maps of operator algebras, *Acta Math.* **110** (1963), 233–278.
42. Varadarajan, V. S., "Geometry of Quantum Theory," Vol. I. Van Nostrand Reinhold, Princeton, New Jersey, 1968.
43. Wigner, E. P., "Group Theory and Its Application to the Quantum Mechanics of Atomic Spectra." Academic Press, New York, 1959.

AMS (MOS) 1980 Subject Classifications: 28B05, 81C20

An Introduction to Nonstandard Analysis and Hyperfinite Probability Theory

PETER A. LOEB

DEPARTMENT OF MATHEMATICS
UNIVERSITY OF ILLINOIS
URBANA, ILLINOIS

I.	Introduction	105
II.	An Introduction to Nonstandard Analysis	108
	A. A Simple Model of the Nonstandard Real Numbers	108
	B. General Nonstandard Models	117
III.	A Nonstandard Representation of Measurable Spaces and L^∞	123
	A. Representations of Measurable Spaces	123
	B. A Representation of L^∞ and its Dual	127
IV.	Conversion from Nonstandard to Standard Measure Spaces	128
	A. Measure Spaces and Integration	128
	B. The Standard Part Map	133
V.	Applications to Stochastic Processes	135
	A. Coin Tossing	135
	B. The Poisson Process	135
	C. Anderson's Construction of Brownian Motion and the Itô Integral	138
	References	141

I. Introduction

It is often helpful in working with stochastic processes to replace them with discrete processes involving infinitesimals. For example, The Poisson process can be thought of as a random distribution of an infinite number of balls into an infinite number of intervals of infinitesimal length. Brownian motion can be thought of as a random walk with infinitesimal steps. In this article, we shall show how Abraham Robinson's nonstandard analysis [24] can be used to make this replacement mathematically rigorous.

Robinson's discovery is applicable to any infinite mathematical structure. For a simple example, we summarize Section II by considering a set-theoretic structure $V(R)$ based on the real numbers R. Each object in $V(R)$ is obtained from the elements of R in a finite number of steps using the usual operations of set theory. For example, the set of all Lebesgue measurable sets is in $V(R)$ as is the set of all Borel measures on R. Let $\mathscr{L}$ be a formal language for $V(R)$; $\mathscr{L}$ contains a name for each object in R: variables, connectives (i.e., $\neg, \vee, \wedge, \rightarrow, \leftarrow$), quantifiers, brackets. The language $\mathscr{L}$ contains also the sentences constructed from these basic symbols. Robinson [24] has shown that there is a (not unique) structure $V(*R)$ built up from a set of individuals $*R$ with the following properties:

(1) Every name of an object in $V(R)$ names something of the same type (i.e., constructed with exactly the same operations from $*R$) in $V(*R)$.

(2) (Transfer Principle) Every sentence in $\mathscr{L}$ that is true for $V(R)$ is true when interpreted in $V(*R)$; quantifiers must, however, be correctly interpreted.

(3) There is an $\eta \in *R$ such that $1 < \eta, 2 < \eta, \ldots$. That is, η is an infinite positive real number.

If A is an object in the structure $V(R)$, then the object in $V(*R)$ with the name A is denoted by $*A$. Either A or $*A$ can be called standard; $*A$ is also called the nonstandard extension of A. For real numbers, we omit the star and think of R as a subset of $*R$.

The transfer principle gives a great deal of information about $V(*R)$. One cannot, however, describe formally what one means by all subsets of an infinite set even when that set is the set N of natural numbers. That is, "all subsets of N" is a primitive notion. If A is a set in R and $\mathscr{P}(A)$ is the power set of A, then $*\mathscr{P}(A) = \mathscr{P}(*A)$ only when A is a finite set. If A is an infinite set, $*\mathscr{P}(A) \subsetneqq \mathscr{P}(*A)$. An element of $*\mathscr{P}(A)$ is called internal, while an element of $\mathscr{P}(*A) - *\mathscr{P}(A)$ is called external. For example, N is external in $*N$. Quantifiers of the form $\forall x \in B$ and $\exists x \in B$ are interpreted in $*R$ to mean $\forall x \in *B$ and $\exists x \in *B$.

The set $*N$ with ordering $* \leqslant$ begins with the standard natural numbers N. (For example, "If $n \in N$ and $n \leqslant 3$, then $n = 0$ or $n = 1$ or $n = 2$ or $n = 3$" is true for N and therefore for $*N$.) After all of the standard elements of N we have an infinite number of elements of $*N - N$ called infinite natural numbers.

The set *R has infinite positive and infinite negative elements. Any other element ρ of *R is infinitely close to a unique standard $r \in R$. That is, $*|\rho - r| \stackrel{*}{\leqslant} 1/n$ for each $n > 0$ in N. We write $\rho \simeq r$ and $r = {}^0\rho$ in this case. A real-valued function f is continuous at $r \in R$ if and only if for each $\rho \simeq r$ in *R, $*f(\rho) \simeq f(r)$. This fact yields a number of properties of continuous functions. For example, if f is continuous on $[0, 1]$, then f takes its maximum value at some point $x_0 \in [0, 1]$. To see this, choose $\eta \in *N - N$ and consider the internal set $\{*f(0), *f(1/\eta), \ldots, *f((\eta - 1)/\eta), *f(1)\}$. Such a set is called *hyperfinite* because it is in internal one-one correspondence with an initial segment of *N and therefore has all of the formal properties of a finite set. In particular, there is a $\gamma \leqslant \eta$ in *N with $*f(\gamma/\eta) \geqslant *f(\omega/\eta)$ for any $\omega \leqslant \eta$ in *N. Let $r_0 = {}^0(\gamma/\eta)$. Then $f(r_0) \geqslant f(r)$ for any $r \in [0, 1]$, since if $r \in [0, 1]$, then there is a greatest $\omega \in *N$ with $\omega/\eta \leqslant r$, whence

$$f(r) \simeq *f(\omega/\eta) \leqslant *f(\gamma/\eta) \simeq f(r_0).$$

Since $f(r)$ and $f(r_0)$ are real, $f(r) \leqslant f(r_0)$.

To illustrate the application of hyperfinite sets in probability theory, we consider an infinite coin toss. Instead of tossing the coin an infinite number of times, we chose some $\gamma \in *N - N$ and consider all of the 2^γ internal ways that we can toss the coin γ times. This nonstandard experiment can be made into a standard measure space, as we shall show, and its rich structure will contain all of the information needed to study standard coin tossing experiments.

We will begin in Section II with an introduction to nonstandard analysis, expanding on the above description of this powerful mathematical method. Our treatment will be a shortened version of the treatment to appear in a forthcoming book by A. E. Hurd and the author. We will not, for example, discuss ultrapowers at length here. The reader who wants to see a detailed construction of a model having the properties we shall describe can now, for example, refer to the book by Stroyan and Luxemburg [27]. We will show in Section III how a standard measure space can be made into a discrete measure space as in [14]. In Section IV we will describe the construction of hyperfinite probability spaces and their application to the weak* limits and cluster points of measures. In Section V we will apply the construction in Section IV to stochastic processes. In particular, we will describe the construction of the Poisson process in [16] and Anderson's construction of Brownian motion and the Itô integral in [1].

Our definitions and results will be numbered consecutively. For example, Theorem II.B.3.5 will be the fifth result or definition in the third part of Section II.B. In Section II.B, however, we will simply write Theorem 3.5.

II. An Introduction to Nonstandard Analysis

A. A Simple Model of the Nonstandard Real Numbers

1. The Real Numbers We consider the standard real numbers first. These form a set R with certain relations that make R a field, a totally ordered set, a metric space, etc.

Definition 1.1 An n-ary relation P on a set S is a subset of the n-fold product $S \times S \times \cdots \times S$ (n factors). If $(a_1, \ldots, a_n) \in P$, we may write $P(a_1, \ldots, a_n)$. A subset B of S defines a unary relation B, and we write either $b \in B$ or $B(b)$ when b is an element of the set B.

In Section A we consider only the set R with all n-ary relations on R and introduce the reader to nonstandard analysis with a simple model for the nonstandard real numbers.

Definition 1.2 We call the set R of real numbers together with all relations on R an elementary structure for the real numbers. We denote this structure by $\mathscr{R}_E$. We will write $a < b, a \leqslant b, a > b, a = b$ instead of $(a, b) \in <, (a, b) \in \leqslant$, etc. If f is a function of n variables on R, then we can write $f(a_1, \ldots, a_n) = b$ for $f(a_1, \ldots, a_n, b)$. We will write $a + b = c$ and $a \cdot b = c$ instead of $+(a, b) = c$ [or $+(a, b, c)$] and $\cdot(a, b) = c$, respectively.

It will be clear from the context whether $f(a, b)$ means "(a, b) satisfies the relation f" or is the name of the image of (a, b) under the function f. Note that $=$ means the subset of R^2 consisting of all pairs where the first and second elements are the same.

2. The Language for $\mathscr{R}_E$ We now introduce a formal language to describe the properties of $\mathscr{R}_E$. We distinguish between the formal statement of a property and the meaning of that statement in $\mathscr{R}_E$ because there are other structures, e.g., hyperreal structures, for which our statements have a meaning and are true when they are true for $\mathscr{R}_E$.

Definition 2.1 A simple language $\mathscr{L}$ for $\mathscr{R}_E$ is a collection of symbols containing:

(a) Names for the objects in $\mathscr{R}_E$, i.e.:

  (i) At least one name $\bar{a}$ for each individual a in R. Such a name is called a constant. Each constant corresponds to at most one individual in R.
  (ii) At least one name $\bar{P}$ for each relation P on R. Each such name is the name of at most one relation on R.

(b) A collection of logical symbols as follows:
   (i) A collection of variables
        $x, x_1, x_2 \ldots y_1, y_2, \ldots$, etc.
   (ii) The connectives $\wedge$ (and) and $\rightarrow$ (implies).
   (iii) The universal quantifier $\forall$ (for all).
   (iv) Parentheses, brackets, comas, and periods.

(c) Terms built by induction according to the following rules:
   (i) Every constant is a term.
   (ii) Every variable is a term.
   (iii) If $\bar{f}$ is the name of a function of n variables on R and $\tau_1, \ldots, \tau_n$ are terms, then $\bar{f}(\tau_1, \ldots, \tau_n)$ is a term.

(d) Sentences (i.e., every variable appears only after a corresponding quantifier) of the form:

(A) $\bar{P}_0(\rho_1, \ldots, \rho_{m_0})$

and

(S) $\forall x_1, \ldots, \forall x_n [\bar{P}_1(\tau_1^1, \ldots, \tau_{m_1}^1) \wedge$
     $\cdots \wedge \bar{P}_k(\tau_1^k, \ldots, \tau_{m_k}^k) \rightarrow \bar{S}_1(\sigma_1^1, \ldots, \sigma_{l_1}^{\ 1}) \wedge \cdots \wedge \bar{S}_j(\sigma_1^j, \ldots, \sigma_{l_j}^j)$

Here, $\bar{P}_i$ denotes an m_i-ary relation on R for $0 \leq i \leq k$, $\bar{S}_i$ denotes an l_i-ary relation on R for $1 \leq i \leq j$, the τ's and σ's are terms involving no variables other than $x_1, \ldots x_n$, and the ρ's are terms involving no variables at all.

We call a sentence such as (A) an atomic sentence and a sentence such as (S) a simple sentence. Their use as the only type of sentence in the language $\mathscr{L}$ and the corresponding use of "Skolem functions" was suggested by a similar use of equalities, inequalities, and Skolem functions in Keisler's work [13]. We shall consider a more general language in Section B.

We have already begun to interpret the language $\mathscr{L}$ in $\mathscr{R}_E$ by assuming knowledge of the objects in $\mathscr{R}_E$ for which the names in $\mathscr{L}$ stand. It follows that terms can be interpreted in $\mathscr{R}_E$.

Definition 2.2 A term is defined in $\mathscr{R}_E$ if it is a constant naming an element of R or if it is of the form $\bar{f}(\sigma_1, \ldots, \sigma_n)$, where each of the terms σ_i contains no variables and is defined, thus naming an element of R, and the n-tuple $(c_1, \ldots, c_n)$ in R^n named by $(\sigma_1, \ldots, \sigma_n)$ is in the domain of the function f named by $\bar{f}$. When defined, $\bar{f}(\sigma_1, \ldots, \sigma_n)$ names that element of R paired with $(c_1, \ldots, c_n)$ in f. If any term in $\bar{f}(\sigma_1, \ldots, \sigma_n)$ is not defined, then $\bar{f}(\sigma_1, \ldots, \sigma_n)$ is not defined.

Definition 2.3 Given the name $\bar{P}$ of an n-ary relation P on R and terms $\tau_1, \ldots, \tau_n$ containing no variables, the atomic sentence

$$\bar{P}(\tau_1, \ldots, \tau_n)$$

is true for $\mathscr{R}_E$ and we write $\mathscr{R}_E \vDash \bar{P}(\tau_1, \ldots, \tau_n)$ if each of the terms $\tau_1, \ldots, \tau_n$ is defined and the n-tuple named by $(\tau_1, \ldots, \tau_n)$ is in P.

We see, for example, that $\overline{\tan}(\overline{\pi/2})$ and $\overline{+}(\overline{1}, \overline{\tan}(\overline{\pi/2}))$ and $\overline{+}(\overline{\ln}(-\overline{3}), \overline{1})$ are undefined while $\overline{\tan}(\overline{\pi/4})$ names 1 and $\overline{+}(\overline{\ln}(\overline{e}), \overline{1})$ names 2. Thus $\overline{=}(\overline{+}(\overline{\ln}(-\overline{3}), \overline{1}), \overline{2})$ and $\overline{=}(\overline{+}(\overline{\ln}(\overline{e}), \overline{1}), \overline{3})$ are not true in $\mathscr{R}_E$ but $\overline{=}(\overline{+}(\overline{\ln}(\overline{e}), \overline{1})\,\overline{2})$ is.

We shall from now on usually abbreviate sentences in $\mathscr{L}$ by omitting some parentheses and the bars from the names; we shall write $+, \cdot, \leq$, etc. in the usual order. For example,

$$\mathscr{R}_E \vDash 1 + \ln e = 2.$$

Definition 2.4 A simple sentence S of the form

$$\forall x_1, \ldots, \forall x_n [P_1(\tau_1^1, \ldots, \tau_{m_1}^1) \wedge \ldots \wedge P_k(\tau_1^k, \ldots, \tau_{m_k}^k) \to$$
$$S_1(\sigma_1^1, \ldots, \sigma_{l_1}^1) \wedge \cdots \wedge S_j(\sigma_1^j, \ldots, \sigma_{l_j}^j)]$$

is true for $\mathscr{R}_E$ and we write $\mathscr{R}_E \vDash S$ if given a replacement of the variables $x_1, \ldots, x_n$ with constants $c_1, \ldots, c_n$ that makes each of the resulting atomic sentences $P_i(\tau_1^i, \ldots, \tau_{m_i}^i)$, $1 \leq i \leq k$, true in $\mathscr{R}_E$, that same replacement makes each of the resulting atomic sentences $S_i(\sigma_1^i, \ldots, \sigma_{l_i}^i)$, $1 \leq i \leq j$, true in $\mathscr{R}_E$.

For example,

$$\forall x [\sqrt{x} > -1 \to \sqrt{x} \geq 0]$$

is true for $\mathscr{R}_E$, but

$$\forall x [R(x) \to \sqrt{x} \geq 0]$$

is not true for $\mathscr{R}_E$ since $\sqrt{}$ is not defined for all real numbers. The complement of an n-ary relation P in R^n is denoted by $\sim P$. If a term in $P(\tau_1, \ldots, \tau_n)$ is not defined, then neither $P(\tau_1, \ldots, \tau_n)$ nor $\sim P(\tau_1, \ldots, \tau_n)$ is true in $\mathscr{R}_E$.

We will abbreviate pairs of simple sentences which are the same except for a reversal of hypothesis and conclusion by writing $\leftrightarrow$. For example, instead of writing $\forall x [x = x \to R(x)]$ and $\forall x [R(x) \to x = x]$, we write $\forall x [x = x \leftrightarrow R(x)]$. Such a sentence is true in $\mathscr{R}_E$ if and only if both of the sentences it replaces are true in $\mathscr{R}_E$.

The fact that R together with $+, \cdot, (\cdot)^{-1}, \leq$ forms an ordered field can be stated with simple sentences. For example,

$$\mathscr{R}_E \vDash \forall x, \forall y [R(x) \wedge R(y) \to x + y = y + x],$$
$$\mathscr{R}_E \vDash \forall x [x \neq 0 \to x \cdot (x)^{-1} = 1].$$

Note that we can use the function $(\cdot)^{-1}$ to replace an existential quantifier. In general, if $\forall x_1, \ldots, \forall x_n, \exists y [P(x_1, \ldots, x_n, y)]$ is true for $\mathscr{R}_E$, then there is a function ϕ of n variables called a *Skolem function* such that the following is true for $\mathscr{R}_E$:

$$\forall x_1, \ldots, \forall x_n [P(x_1, \ldots, x_n, \phi(x_1, \ldots, x_n))].$$

For example, if f is a function of one variable with domain $A \subset R$ and range $B \subset R$, then the fact that B is the range of f is expressed using a Skolem function ϕ as follows:

$$\forall x [f(x) = f(x) \to B(f(x))],$$
$$\forall y [B(y) \to f(\phi(y)) = y].$$

The range of a function of n variables can be similarly characterized. The fact that A is the domain of the above function f can be expressed as follows:

$$\forall x [A(x) \leftrightarrow f(x) = f(x)].$$

B. The Hyperreal Numbers We shall now imbed R in an ordered field $*R$ whose elements are called nonstandard real numbers or hyperreal numbers. There is not a unique nonstandard extension of R, but any nonstandard extension yields a great deal of information about R.

To be exact, we consider a structure $*\mathscr{R}_E$ consisting of a set $*R$ and some relations on $*R$ together with a mapping $*(\cdot)$ from $\mathscr{R}_E$ into $*\mathscr{R}_E$ so that the following is true:

(i) If $c \in R$, then $*(c) \in *R$. If $c \in R$ and $d \in R$ and $c \neq d$, then $*(c) \neq *(d)$.
(ii) If P is an n-ary relation on R, then $*(P)$ is an n-ary relation on $*R$.
(iii) If f is a function of n variables on R, then $*(f)$ is a function of n variables on $*R$.
(iv) The relation $*(=)$ is the equality relation on $*R$, and $*(\neq)$ is its complement in $*R \times *R$.

We extend the language $\mathscr{L}$ so that it becomes a language for $*\mathscr{R}_E$ as follows:

(i) The name $\bar{c}$ of an individual c in R is the name of $*(c)$ in $*R$.
(ii) The name $\bar{P}$ of a relation P on R is the name of $*(P)$ on $*R$.

(iii) We adjoin to $\mathscr{L}$ at least one name for each individual not yet named by (i). No name in $\mathscr{L}$ names more than one individual in $*R$. The names of individuals in $*R$ are called constants.

(iv) Having new constants in $\mathscr{L}$, we adjoin all possible new terms of the form $\bar{f}(\tau_1, \ldots, \tau_n)$, where $\bar{f}$ names a function f of n variables on R (and a function $*(f)$ of n variables on $*R$) and $\tau_1, \ldots, \tau_n$ are terms in $\mathscr{L}$.

(v) We adjoin to $\mathscr{L}$ all possible new atomic sentences of the form $\bar{P}(\tau_1, \ldots, \tau_n)$, where $\bar{P}$ is the name of an n-ary relation P on R (and an n-ary relation $*(P)$ on $*R$) and $\tau_1, \ldots, \tau_n$ are terms in $\mathscr{L}$ containing no variables.

The notion of a term in $\mathscr{L}$ being defined or not defined in $*\mathscr{R}_E$ is the same as in Definition 2.2 except that we replace R with $*R$ and $\mathscr{R}_E$ with $*\mathscr{R}_E$. Similarly, an atomic sentence $\bar{P}(\tau_1, \ldots, \tau_n)$ is true for $*\mathscr{R}_E$ and we write $*\mathscr{R}_E \vDash \bar{P}(\tau_1, \ldots, \tau_n)$ if each term τ_i is defined in $*\mathscr{R}_E$ and the n-tuple in $(*R)^n$ named by $(\tau_1, \ldots, \tau_n)$ is in $*(P)$. We have not augmented the collection of simple sentences in $\mathscr{L}$.

Definition 3.1 A simple sentence S of the form

$$\forall x_1, \ldots, \forall x_n [P_1(\tau_1^1, \ldots, \tau_{m_1}^1) \wedge \cdots \wedge P_k(\tau_1^k, \ldots, \tau_{m_k}^k) \to$$
$$S_1(\sigma_1^1, \ldots, \sigma_{l_1}^1) \wedge \cdots \wedge S_j(\sigma_1^j, \ldots, \sigma_{l_j}^j)]$$

is true for $*\mathscr{R}_E$, and we write $*\mathscr{R}_E \vDash S$, if for each replacement of the variables $x_1, \ldots, x_n$ with constants $c_1, \ldots, c_n$ such that all of the atomic sentences $P_i(\tau_1^1, \ldots, \tau_{m_i}^1)$ are true for $*\mathscr{R}_E$, the atomic sentences $S_i(\sigma_1^i, \ldots, \sigma_{l_i}^i)$ are all true for $*\mathscr{R}_E$.

Definition 3.2 We call $*\mathscr{R}_E$ a simple hyperreal structure or a simple nonstandard extension of $\mathscr{R}_E$, if

(i) There is an element γ of $*R$ such that for each c in R, $(*(c), \gamma)$ satisfies the relation $*(\leq)$. (For example, $*1 * \leq \gamma$, $*2 * \leq \gamma$, $*3 * \leq \gamma$, etc.)

(ii) (Simple Transfer Principle) Every simple sentence in $\mathscr{L}$ that is true for $\mathscr{R}_E$ is true for $*\mathscr{R}_E$.

It is usually the transfer principle rather than any particular construction that should be kept in mind in working with nonstandard analysis. A construction can be obtained starting with a free ultrafilter $\mathscr{U}$ in N. That is, $\phi \notin \mathscr{U}$; if $A \in \mathscr{U}$ and $B \in \mathscr{U}$, then $A \cap B \in \mathscr{U}$; for each set $A \subset N$, either $A \in \mathscr{U}$ or $N - A \in \mathscr{U}$, and $\{n\} \notin \mathscr{U}$ for any $n \in N$. Two sequences $\{r_n\}$ and $\{s_n\}$ in R are called equivalent if $\{n \in N : r_n = s_n\} \in \mathscr{U}$. We write $[r_n]$ for the equivalence class containing $\{r_n\}$; these equivalence classes are the elements of $*R$. The mapping $*(\cdot)$ sends each $r \in R$ to $[r_n] \in *R$, where $r_n = r$ for all n. An n-tuple $([r_m^1], \ldots, [r_m^n])$ is in $*(P)$, where P is an n-ary relation on R, if and only if $\{k \in N : (r_k^1, \ldots, r_k^n) \in P\} \in \mathscr{U}$. If $s_n = n$ for each $n \in N$, then

[s_n] is an infinite integer. The reader is referred to [13] or [27] for further details.

When $c \in R$, we shall henceforth write c instead of $*(c)$ and think of c as both an element of R and of $*R$. If P is an n-ary relation on R and the n-tuple $(c_1, \ldots, c_n)$ of real numbers is in P, then

$$\forall x[x = x \to P(c_1, \ldots, c_n)]$$

is true for $\mathscr{R}_E$ and $*\mathscr{R}_E$. Thus, P, considered as a set of n-tuples, is a subset of $*(P)$. In particular, if f is an $(n + 1)$-ary relation which is a function of its first n arguments, then $*(f)$ extends the function f.

The relation $\sim P$ is the complement of an n-ary relation P in R^n. The characteristic function of P is the function χ_P of n variables on R^n such that $\chi_P(c_1, \ldots, c_n) = 1$ if $(c_1, \ldots, c_n) \in P$ and $\chi_P(c_1, \ldots, c_n) = 0$ if $(c_1, \ldots, c_n) \in \sim P$. The sentences

$$\forall x_1, \ldots, \forall x_n[5 = 5 \to R(\chi_P(x_1, \ldots, x_n))],$$
$$\forall x_1, \ldots, \forall x_n[\chi_P(x_1, \ldots, x_n) \neq 0 \to \chi_P(x_1, \ldots, x_n) = 1],$$
$$\forall x_1, \ldots, \forall x_n[P(x_1, \ldots, x_n) \leftrightarrow \chi_P(x_1, \ldots, x_n) = 1],$$
$$\forall x_1, \ldots, \forall x_n[\sim P(x_1, \ldots, x_n) \leftrightarrow \chi_P(x_1, \ldots, x_n) = 0]$$

are true for $\mathscr{R}_E$ and therefore $*\mathscr{R}_E$. Thus $*(\chi_P)$ is the characteristic function on $(*R)^n$ of $*(P)$ and $*(\sim P) = \sim(*(P))$.

Definition 3.3 If A is a subset of R and χ_A is its characteristic function, then the nonstandard extension $*A$ of A is the set $\{s \in *R : *(\chi_A)(s) = 1\}$.

Note that for a set A in R, $*\mathscr{R}_E \vDash A(c)$ if and only if $c \in *A$. We write $*A$ instead of $*(A)$ for the unary relation A. Similarly, for any function f or relation P in $\mathscr{R}_E$, we write $*f$ and $*P$ instead of $*(f)$ and $*(P)$. Thus we now distinguish between a relation P, its name P (or $\bar{P}$) in $\mathscr{L}$, and the corresponding relation $*P$ in $*\mathscr{R}_E$. Note that $\chi_{*P} = *\chi_P$. We do not omit the star (except for $\leq, <, \geq, >, +, \cdot, (\cdot)^{-1}$, and $|\cdot|$) since $*P$ may be a proper extension of P. For functions we have the following result:

Theorem 3.4 *Let f be a function of n variables in $\mathscr{R}_E$ with domain $D_f \subset R^n$ and range $I_f \subset R$. Then $(a_1, \ldots, a_n) \in *D_f$ if and only if $(a_1, \ldots, a_n)$ is in the domain of $*f$, and $b \in *I_f$ if and only if b is in the range of $*f$. If $(a_1, \ldots, a_n) \in D_f$, then $*f(a_1, \ldots, a_n)$ is defined and equals $f(a_1, \ldots, a_n)$.*

Proof The theorem follows from the fact that the domain and range of f can be characterized by simple sentences using Skolem functions.

Again using simple sentences we can obtain the following result which justifies omitting the stars from $+, \leq$, etc.

Theorem 3.5 *The set *R together with $+$, $\cdot$, $(\cdot)^{-1}$, and $\leqslant$ is a proper ordered field extension of $(R, +, \cdot, (\cdot)^{-1}, \leqslant)$. The extension of $|\cdot|$ is the absolute value function on *R.*

If A is a set in R^n, then *A will have elements not in A if and only if A is infinite. We call the elements of A standard elements of *A.

Definition 3.6 An element $\alpha \in$ *R is called an infinite positive hyperreal number (or nonstandard real number) if $\alpha > r$ for each $r \in R$; $\alpha \in$ *R is called an infinite negative hyperreal number if $\alpha < r$ for each $r \in R$. An element $\alpha \in$ *R that is not infinite is called a finite hyperreal (or nonstandard real) number. Such an $\alpha \in$ *R is called infinitesimal if $|\alpha| < r$ for all $r > 0$ in R, and we write $\alpha \simeq 0$. If α and β are in *R and $\alpha - \beta \simeq 0$, we write $\alpha \simeq \beta$. If $\alpha \simeq \beta$ and β is in R, we write $\beta = {}^0\alpha$ and called β the standard part of α. For a positive infinite α we write $+\infty = {}^0\alpha$, and for a negative infinite α we write $-\infty = {}^0\alpha$.

Using Dedekind cuts, it is easy to see that for a finite $\alpha \in$ *R there is one and only one $r \in R$ with $\alpha \simeq r$. The set $m(r) = \{\alpha \in {}^*R : \alpha \simeq r\}$ is called the monad of r.

Theorem 3.7 *If α and β are finite hyperreal numbers and $r \in R$, then the following is true:*

(i) ${}^0(\alpha + \beta) = {}^0\alpha + {}^0\beta$,   (ii) ${}^0(\alpha - \beta) = {}^0\alpha - {}^0\beta$,

(iii) ${}^0(\alpha \cdot \beta) = {}^0\alpha \cdot {}^0\beta$,   (iv) ${}^0(\alpha/\beta) = {}^0\alpha/{}^0\beta$ if ${}^0\beta \neq 0$,

(v) ${}^0r = r$,   (vi) $\alpha \simeq \beta \to {}^0\alpha = {}^0\beta$,

(vii) $\alpha \leqslant \beta \to {}^0\alpha \leqslant {}^0\beta$.

Proof We will prove (iii). If $\alpha = r + \varepsilon$ and $\beta = s + \delta$, where $r \in R$, $s \in R$, $\varepsilon \simeq 0$, and $\delta \simeq 0$, then $(r + \varepsilon)(s + \delta) = rs + r\delta + s\varepsilon + \varepsilon\delta \simeq rs$, since $|r\delta| \leqslant |r| \cdot (1/n)$ for each $n \in N$, i.e., $r\delta \simeq 0$, $s\varepsilon \simeq 0$, and $\varepsilon\delta \simeq 0$.

4. Applications to Real Analysis We now give a sample of results which show some of the connections between *$\mathscr{R}$ and $\mathscr{R}$ for any hyperreal structure *$\mathscr{R}$. These results are originally due to Robinson [24]. We will assume in this section that $0 \notin N$.

Theorem 4.1 *A set $A \subset R$ is open if and only if for each $r \in A$, $m(r) \subset {}^*A$.*

Proof If A is open and $r \in A$, then for some $\delta > 0$ in R, "$\forall x[|x - r| < \delta \to A(x)]$" is true for $\mathscr{R}_E$ and therefore *$\mathscr{R}_E$. If $\alpha \in {}^*R$ and $\alpha \simeq r$, then $|\alpha - r| < \delta$, whence $\alpha \in {}^*A$. Assume now that there is an $r \in A$ such that to each $n \in N$ there corresponds a point $\phi(n) \in R$ with $|r - \phi(n)| < 1/n$ and $\phi(n) \notin A$. Then

$$\forall x[N(x) \to |r - \phi(x)| < (1/x) \land \sim A(\phi(x))]$$

is true for $*\mathscr{R}_E$. Thus if $\omega \in *N - N$, we have $*\phi(\omega) \simeq r$ but $*\phi(\omega) \notin *A$, i.e., $m(r) \not\subset *A$. It now follows that if $m(r) \subset *A$ for each $r \in A$, then for each $r \in A$ there is an $n_r \in N$ such that $s \in A$ whenever $|s - r| < 1/n_r$.

Corollary 4.2 *A set $A \subset R$ is closed if and only if for each $r \in R$ we have $r \in A$ whenever $m(r) \cap *A \neq \varnothing$.*

Recall that a set in a topological space is compact if and only if each open covering has a finite subcovering.

Theorem 4.3 *A set $A \subset R$ is compact if and only if each $\rho \in *A$ is finite and $^0\rho \in A$.*

Proof If A is compact and there is a $\rho_0 \in *A$ that is not in the monad of any $r \in A$, then for each $r \in A$ there is a $\delta_r > 0$ in R such that $|\rho_0 - r| \geq \delta_r$. Since A is compact, we can find a finite number of points $r_i \in A$, $1 \leq i \leq n$, such that the following is true for $\mathscr{R}_E$ and therefore $*\mathscr{R}_E$:

$$\forall y[A(y) \wedge |r_1 - y| \geq \delta_{r_1} \wedge \cdots \wedge |r_{n-1} - y| \geq \delta_{r_{n-1}} \to |r_n - y| < \delta_{r_n}].$$

But $\rho_0 \in *A$ and $|r_i - \rho_0| \geq \delta_{r_i}$ for all $i \in N$, $i \leq n$. It follows that for a compact A, each $\rho \in *A$ is in the monad of some $r \in A$.

Conversely, assume $A \subset R$ and there is an open covering $\mathcal{O}$ of A with no finite subcovering. We may assume that $\mathcal{O}$ consists of a countable number of intervals of the form $(f(n) - g(n), f(n) + g(n))$, $n \in N$. For each $n \in N$, there is a point $h(n) \in A$ such that $|h(n) - f(j)| \geq g(j) > 0$ for all $j \leq n$ in N. The functions $f, g,$ and h have extensions to $*\mathscr{R}_E$, and

$$*\mathscr{R}_E \vDash \forall x, \forall y[N(x) \wedge N(y) \wedge y \leq x \to A(h(x)) \wedge |h(x) - f(y)| \geq g(y)].$$

Thus, if $\omega \in *N - N$, then $*h(\omega) \in *A$, but for each $r \in A$, $*h(\omega) \not\simeq r$ since, for some $n \in N$ and $\varepsilon > 0$ in R, $|f(n) - r| \leq g(n) - \varepsilon$ while $|f(n) - *h(\omega)| \geq g(n)$. Therefore, if each $\rho \in *A$ is in the monad of some $r \in A$, then A is compact.

Corollary 4.4 (Heine–Borel) *A set $A \subset R$ is compact if and only if it is closed and bounded.*

Theorem 4.5 *A real-valued function f with domain $D_f \subset R$ is continuous at a point $r \in D_f$ if and only if for each $\rho \simeq r$ in $*D_f$ we have $*f(\rho) \simeq f(r)$. The function f is uniformly continuous on D_f if and only if for each pair of points α and β in $*D_f$ with $\alpha \sim \beta$ we have $*f(\alpha) \simeq *f(\beta)$.*

Proof Assume f is continuous at $r \in D_f$, and recall that $*f(r) = f(r)$. Fix $\rho \in *D_f$ with $\rho \simeq r$. We will show that $|*f(\rho) - *f(r)| < \varepsilon$ for every $\varepsilon > 0$ in R. Given $\varepsilon > 0$ in R, there is a $\delta > 0$ in R such that the following is true for $\mathscr{R}_E$ and $*\mathscr{R}_E$:

$$\forall y[D_f(y) \wedge |y - r| < \delta \to |f(y) - f(r)| < \varepsilon].$$

Since $\rho \in {}^*D_f$ and $|\rho - r| < \delta$, $|{}^*f(\rho) - {}^*f(r)| < \varepsilon$. Conversely, assume that there is an $\varepsilon > 0$ in R such that for each $n \in N$ there is a point $\phi(n) \in D_f$ with $|\phi(n) - r| < 1/n$ and $|f(\phi(n)) - f(r)| \geq \varepsilon$. Then

$${}^*\mathcal{R}_E \vDash \forall x[N(x) \to D_f(\phi(x)) \wedge |\phi(x) - r| < (1/x) \wedge |f(\phi(x)) - f(r)| \geq \varepsilon].$$

For $\omega \in {}^*N - N$ we have ${}^*\phi(\omega) \in {}^*D_f$ and ${}^*\phi(\omega) \simeq r$, but ${}^*f({}^*\phi(\omega)) \not\simeq f(r)$. The rest of the proof is left to the reader.

Corollary 4.6 *A real-valued function f that is continuous on a compact set $A \subset R$ is uniformly continuous on A.*

Theorem 4.7 (Intermediate Value Theorem) *If f is a continuous real-valued function on $[0, 1]$ and $f(0) < 0$ while $f(1) > 0$, then there is a point $c \in (0, 1)$ with $f(c) = 0$.*

Proof Let g be the function on N such that for each $n \in N$, $g(n) \in N$, $g(n) \leq n$, $f(g(n)/n) \geq 0$, and $f(m/n) < 0$ for all $m \in N$ with $m < g(n)$. Then for $\omega \in {}^*N - N$,

$${}^*f\left(\frac{{}^*g(\omega)}{\omega}\right) \geq 0, \quad {}^*f\left(\frac{{}^*g(\omega) - 1}{\omega}\right) < 0,$$

and

$${}^*f\left(\frac{{}^*g(\omega)}{\omega}\right) \simeq {}^*f\left(\frac{{}^*g(\omega) - 1}{\omega}\right) \simeq f\left({}^0\!\left(\frac{{}^*g(\omega)}{\omega}\right)\right),$$

whence $f({}^0({}^*g(\omega)/\omega)) = 0$.

Theorem 4.8 *If f is a real-valued function defined on R then f has limit L at $a \in R$ if and only if for each $\rho \simeq a$ with $\rho \neq a$ we have ${}^*f(\rho) \simeq L$.*

Proof The proof is left to the reader.

Statements similar to Theorem 4.8 can be made for $\lim_{x \to +\infty} f(x) = L$, $\lim_{x \to a} f(x) = +\infty$, etc. Also, of course, corresponding statements can be made when the domain of f is a proper subset of R.

A sequence $\{r_n\}$ in R is a function from N into R; it has an extension $\{\rho_n : n \in {}^*N\}$ that maps ${}^*N$ into ${}^*R$ so that for each $n \in N$, $\rho_n = r_n$. We write r_n for the nth element of the sequence even when $n \in {}^*N - N$.

Theorem 4.9 *If $\{r_n\}$ is a sequence in R, then $c \in R$ is a cluster point of $\{r_n\}$ if and only if there is some $\omega \in {}^*N - N$ such that $r_\omega \simeq c$. The point c is a limit of $\{r_n\}$ if and only if for all $\omega \in {}^*N - N$, $r_\omega \simeq c$.*

Proof The proof is left to the reader.

Corollary 4.10 (Bolzano–Weierstrass) *A bounded infinite sequence $\{r_n\}$ in R has a cluster point.*

Integrals and derivatives can, of course, be defined using infinitesimals. If f is defined on an open interval containing x_0, then the derivative of f exists at x_0 and equals b if and only if for all $\rho \simeq x_0$ with $\rho \neq x_0$ we have $[{}^*f(\rho) - {}^*f(x_0)]/(\rho - x_0) \simeq b$. For example, if $f(x) = x^2$ for all $x \in R$, then for $\Delta x \neq 0$ but $\Delta x \simeq 0$,

$$[{}^*f(x + \Delta x) - f(x)]/\Delta x = [x^2 + 2x\Delta x + (\Delta x)^2 - x^2]/\Delta x = 2x + \Delta x \simeq 2x.$$

If f is continuous on $[a, b]$, we let $S_f(n) = \sum_{k=1}^{n} f(a + (k/n)(b - a)) \cdot (b - a)/n$. For $\omega \in {}^*N - N$, ${}^*S_f(\omega) \simeq \int_a^b f(x)\,dx$.

B. General Nonstandard Models

1. Superstructures In order to work with analysis beyond calculus, we need to consider sets of sets, sets of functions, etc. We will also use a richer language than that employed in Section A. We start with a set S and all of the sets one can construct from S in a finite number of steps.

Definition 1.1 The power set $\mathscr{P}(S)$ of a set S is the collection of all subsets of S. The nth cumulative power set of S is defined by recursion as follows:

$$V_0(S) = S, \quad V_{n+1}(S) = V_n(S) \cup \mathscr{P}(V_n(S))$$

The superstructure over S, $V(S)$, is the union

$$V(S) = \bigcup_{n=0}^{\infty} V_n(S).$$

We start with a set S of individuals or atoms; i.e., if $s \in S$, then s has no members. We assume that the empty set is not an individual in S but each real number is. Each object in $V(S)$ is obtained from S in a finite number of steps.

If a and b are in $V_n(S)$, then $(a, b) = \{\{a\}, \{a, b\}\}$ is in $V_{n+2}(S)$. In general, an n-tuple

$$(a_1, a_2, \ldots, a_n) = \{\{a_1\}, \{a_1, a_2\}, \{a_1, a_2, a_3\}, \ldots, \{a_1, \ldots, a_n\}\}$$

of elements of $V_n(S)$ is in $V_{n+2}(S)$. Thus an n-ary relation P on $V_n(S)$ is in $V_{n+3}(S)$. The relation P is also an n-ary relation on $V_{n+m}(S)$ for $m \geq 1$. On the other hand, $\sim P$ on $V_n(S)$ is a proper subset of $\sim P$ on $V_{n+m}(S)$ when $m \geq 1$. For example, $\sim <$ on R does not contain the pair $(5, R)$. If P is an n-ary relation on $V_n(S)$ but not on $V_{n-1}(S)$, we call n the level of P. We let $(=)_n, (\in)_n, (\neq)_n$, and $(\notin)_n$ denote $=, \in, \sim =$, and $\sim \in$ on $V_n(S)$.

We now consider a formal language $\mathscr{L}_S$ for $V(S)$. $\mathscr{L}_S$ contains the following:

(i) A symbol, called a constant, for each element of $V(S)$. A constant names at most one element of $V(S)$.

(ii) Variables.
(iii) The connectives $\neg, \wedge, \vee, \rightarrow, \leftarrow, \leftrightarrow$.
(iv) The quantifiers $\forall, \exists$.
(v) The relation symbols $\in$ and $=$.
(vi) Parentheses, commas, and periods.
(viii) Terms formed by induction so that

(a) every constant is a term;
(b) every variable is a term;
(c) if $\tau_1, \ldots, \tau_n$ are terms and f is the name of a function f of n variables in $V(S)$ [whence $f \in V_m(S)$ for some $m \in N$], then $f(\tau_1, \ldots, \tau_n)$ is a term. A term with no variables is called a closed term.

(viii) Formulas defined by induction so that

(a) if τ_1 and τ_2 are terms, then $\tau_1 = \tau_2$ and $\tau_1 \in \tau_2$ are formulas, called atomic formulas, of $\mathscr{L}_S$;
(b) if $\tau_1, \ldots, \tau_n$ are terms and P is the name of an n-ary relation in $V(S)$ [whence $P \in V_m(S)$ for some $m \in N$], then $P(\tau_1, \ldots, \tau_n)$ is a formula, also called an atomic formula, of $\mathscr{L}_S$;
(c) if ϕ and ψ are formulas of $\mathscr{L}_S$, so are $\neg \phi, \phi \wedge \psi, \phi \vee \psi, \phi \rightarrow \psi, \psi \rightarrow \phi$ and $\phi \leftrightarrow \psi$;
(d) if x is a variable and A is a constant naming a nonempty set in $V_m(S)$ for some $m \geq 1$, and if ϕ is a formula containing no formula of the form $(\forall x \in B)\psi$ or $(\exists x \in B)\psi$, then $(\forall x \in A)\phi$ and $(\exists x \in A)\phi$ are formulas of $\mathscr{L}_S$.

Definition 1.2 A sentence of $\mathscr{L}_S$ is a formula in which every occurrence of a variable is within the scope of a quantifier.

If $A \in V_{m+1}(S)$ is a nonempty set of functions of one variable with $f \in A$ and $f(a) = b$, we cannot say $(\exists x \in A)[x(a) = b]$. There is, however, a function T_m of two variables defined on $V_m(S)$ consisting of pairs $((g, c), d)$, where g is a function, c is in the domain of g, and $d = g(c)$. Given A, f, a, and b as above, we can say

$$(\exists x \in A)[T_m(x, a) = b].$$

Also, we cannot say

$$(\exists y \in V_{m+1}(S))((\exists x \in y)[T_m(x, a) = b]).$$

We can say

$$(\exists y \in V_{m+1}(S))((\exists x \in V_m(S))[x \in y \wedge T_m(x, a) = b]).$$

Definition 1.3 A term of $\mathscr{L}_S$ is defined in $V(S)$ if it is a constant naming an element of S or a term of the form $f(\tau_1, \ldots, \tau_n)$, where f names a function

f of n variables in $V(S)$, each of the terms τ_i is closed and defined in $V(S)$, and the n-tuple named by $(\tau_1, \ldots, \tau_n)$ is in the domain of f. An atomic sentence $\tau_1 = \tau_2$ (or $\tau_1 \in \tau_2$) is true in $V(S)$ if τ_1 and τ_2 are closed terms defined in $V(S)$ and the element in $V(S)$ named by τ_1 is equal to (is an element of) the element of $V(S)$ named by τ_2. An atomic sentence $P(\tau_1, \ldots, \tau_n)$ is true in $V(S)$ if the terms $\tau_1, \ldots, \tau_n$ are closed and defined in $V(S)$ and the n-tuple named is in the set named by P. A sentence of the form $(\forall x \in A)\psi$ is true in $V(S)$ if each replacement of the variable x with a name of an element of the nonempty set A makes the resulting sentence ψ true in $V(S)$. A sentence of the form $(\exists x \in A)\psi$ is true in $V(S)$ if there is some replacement of the variable x with a name of an element of A that makes the resulting sentence ψ true in $V(S)$. Let ϕ and ψ be sentences of $\mathscr{L}_S$. If ϕ and ψ are both true in $V(S)$, then $\phi \wedge \psi$, $\phi \vee \psi$, $\phi \rightarrow \psi$, $\psi \rightarrow \phi$, and $\phi \leftrightarrow \psi$ are true in $V(S)$, while $\neg \phi$ and $\neg \psi$ are not true in $V(S)$. If ϕ is true in $V(S)$ and ψ is not true in $V(S)$, then $\phi \vee \psi$, $\neg \psi$, and $\psi \rightarrow \phi$ are true in $V(S)$ but $\phi \wedge \psi$, $\neg \phi$ $\phi \rightarrow \psi$ and $\phi \leftrightarrow \psi$ are not true in $V(S)$. If neither ϕ nor ψ is true in $V(S)$, then $\neg \phi$, $\neg \psi$, $\phi \rightarrow \psi$, $\psi \rightarrow \phi$, and $\phi \leftrightarrow \psi$ are true in $V(S)$, while $\phi \vee \psi$ and $\phi \wedge \psi$ are not true in $V(S)$. If a sentence ψ in $\mathscr{L}_S$ is true in $V(S)$, we write $V(S) \vDash \psi$.

Note that $\neg[\sqrt{-3} = 2]$ is true in $V(S)$, but $\sqrt{-3} \, (\neq)_0 \, 2$ is not true in $(V(S)$.

2. Nonstandard Structures Given S and the structure $V(S)$, we now consider a set $*S$ and the corresponding structure $V(*S)$ together with a map $*$ from $V(S)$ into $V(*S)$. We assume that the map $*$ has the properties in the following list:

(i) For each $s \in S$, $*(s)$ is in $*S$, and moreoever, $*(S) = *S$.
(ii) For any $m \in N$ and $A \in V_{m+1}(S) - V_m(S)$, $*(A)$ is in $V_{m+1}(*S) - V_m(*S)$.
(iii) For any $m \in N$,

$$*(=)_m \quad \text{is } (=)_m \text{ on } V_m(*S),$$
$$*(\neq)_m \quad \text{is } (\neq)_m \text{ on } V_m(*S),$$
$$*(\in)_m \quad \text{is } (\in)_m \text{ on } V_m(*S),$$
$$*(\notin)_m \quad \text{is } (\notin)_m \text{ on } V_m(*S).$$

(iv) Given A and B in $V_m(S)$, $*(A) = *(B)$ if and only if $A = B$, and $*(A) \in *(B)$ if and only if $A \in B$.

(v) If P is an n-ary relation on $V_m(S)$ for some m, then $*(P)$ is an n-ary relation on $V_m(*S)$. If f is a function of n variables on $V_m(S)$, that is, if the domain of f consists of n-tuples with entries from $V_m(S)$, then $*(f)$ is a function of n variables on $V_m(*S)$.

(vi) If $B \in *(V_m(S))$ for some $m \in N$ and $A \in B$, then $A \in (*V_{m-1}(S))$.

Note that we do not have the property $*(V_m(S)) = V_m(*S)$ except for $m = 0$. We shall hereafter usually omit parentheses in writing $*(A)$ and write $*A$ instead. If s is an individual in S, then we shall write s for both s and $*s$.

To form a language for $V(*S)$, we let each constant in $\mathscr{L}_S$ name both an object in $V(S)$ and the image of that object under the mapping $*$. We adjoin at least one constant as a name for each object in $V(*S)$ that is not the image of something under the mapping $*$. No constant names more than one element of $V(*S)$. With this choice of constants, we continue to form a language for $V(*S)$ as we would for any superstructure. We denote the language formed in this way $*\mathscr{L}_S$. A sentence in $\mathscr{L}_S$ is also a sentence, called a standard sentence, in $*\mathscr{L}_S$.

Definition 2.1 Given a set A in $V(S)$, we let $\mathscr{P}_F(A)$ denote the collection of all finite subsets of A.

Definition 2.2 A superstructure $V(*S)$ with the mapping $*$ is called an enlargement of $V(S)$ if every sentence in $\mathscr{L}_S$ that is true for $V(S)$ is true for $V(*S)$ and if for each set A in $V(S)$ there is a set $B \in *\mathscr{P}_F(A)$ such that $*a \in B$ for each $a \in A$.

If A is a set in $V(S)$ and $B \in *\mathscr{P}_F(A)$, then B has all of the formal properties of a finite set. Such a set B is called a hyperfinite set.

Although we shall not give a proof of the existence of enlargements here, we can give some indications of a construction by considering a free ultra filter $\mathscr{U}$ on the natural numbers N. (The set N is too small for the general construction.) The extension ($*R$ of R consists, as in Section A, of equivalence classes $[r_n]$ formed from sequences $\{r_n\}$ of real numbers. Recall that $\{r_n\}$ is equivalent to $\{s_n\}$ if $\{n \in N : r_n = s_n\} \in \mathscr{U}$. We say that two sequences $\{A_n\}$ and $\{B_n\}$ of subsets of R are equivalent if $\{n \in N : A_n = B_n\} \in \mathscr{U}$. We let $[A_n]$ denote the equivalence class containing $\{A_n\}$. We say that $[r_n] \in' [A_n]$ if $\{n \in N : r_n \in A_n\} \in \mathscr{U}$. A set $\tilde{A} \subset *R$ is called an internal set if there is an equivalence class $[A_n]$ such that $[r_n] \in \tilde{A}$ if and only if $[r_n] \in' [A_n]$. Any subset of $*R$ that is not internal is called external. It is easy to show that N is external in $*R$.

If $A \subset R$, then $*A$ corresponds to the equivalence class $[A_n]$, where $A_n = A$ for all n. That is, $[r_n] \in *A$ if and only if $\{n \in N : r_n \in A\} \in \mathscr{U}$. When $A \subset R$, $*A$ is called the nonstandard extension of A. If $B_n = \{1, 2, \ldots, n\}$ for all $n \in N$, then the set $\tilde{B}$ corresponding to $[B_n]$ is a hyperfinite set with $n \in \tilde{B}$ for each $n \in N$. For the general case, we have the following definition.

Definition 2.3 If $A \in V(S)$, then we shall say that A or $*A$ is standard and that $*A$ is the nonstandard extension of A. If $B \in V(*S)$, then we call B internal if $B \in *A$ for some standard A, in particular, if $B \in *V_m(S)$ for some

$m \in N$. An object B in $V(*S)$ that is not internal is called external. Every individual in $*S$ is internal.

Fix a set S containing R, and consider $V(S)$ and $V(*S)$. As before, we write $\leq$, $+$, etc., instead of $*\leq$, $*+$, etc.

Proposition 2.4 *There is an infinite element $\eta \in *N$. That is, $n \leq \eta$ for each standard $n \in N$.*

Proof Let Max denote the function on $\mathscr{P}_F(N)$ which chooses the maximum element for each finite subset of N. Let $B \in *\mathscr{P}_F(N)$ be a hyperfinite set with $n \in B$ for each $n \in N$ and let $\eta = \text{Max}(B)$. Then $n \leq \eta$ for each $n \in N$ since $(\forall x \in \mathscr{P}_F(N))(\forall y \in N)[y \in x \to y \leq \text{Max}(x)]$ is true for $V(S)$ and $V(*S)$.

Proposition 2.5 *The set N is external in $V(*S)$.*

Proof Assume that N is internal. Then $*N - N$ is also internal since
$$(\forall x \in V_1(S) - S)(\forall y \in V_1(S) - S)[y - x \in V_1(S)]$$
is true in $V(*S)$. Here, of course, $-$ denotes the operation on sets and $V_1(S)$ and S are interpreted in $V(*S)$ as meaning $*V_1(S)$ and $*S$, respectively. Since $*N - N$ is not empty and, we assume, internal, there is a first element ω. Since $\omega - 1 \in N$, $\omega = (\omega - 1) + 1 \in N$; but this is impossible.

When one says "all subsets of N," there is no formal way of saying what one means. In $V(*S)$, we interpret $(\forall x \in \mathscr{P}(N))$ as meaning $\forall x \in *\mathscr{P}(N)$; again, $*\mathscr{P}(N) \subsetneq \mathscr{P}(*N)$. In general, "$\forall x \in A$" and "$\exists x \in A$" are interpreted in $V(*S)$ as meaning "for all x in $*A$" and "there exists an x in $*A$," respectively.

We have said that A is internal in $V(*S)$ if $A \in *B$ for some standard B in $V(S)$. Clearly, the nonstandard extension $*B$ of a standard B in $V(S)$ is internal in $V(*S)$. Moreover, by property (vi) of the mapping $*$, if B is internal in $V(*S)$ and $A \in B$, then A is internal. We also have the following internal definition principle.

Theorem 2.6 [Keisler [13]] *Let $\phi(x)$ be a formula in $*\mathscr{L}_S$ for which x is the only variable not within the scope of a quantifier. Assume that each name in $\phi(x)$ is the name of an internal element of $V(*S)$. Let A be an internal set in $V(*S)$. Then $\{a \in A : V(*S) \vDash \phi(a)\}$ is an internal set in $V(*S)$.*

Proof We may assume that $\phi(x)$ has been written so that the only relation symbols and function symbols used are names of standard relations and functions. [For example, if P is a binary relation in $*V_{m+3}(S)$ and the pair $c = (a, b)$ is in P, then letting U_1^{m+2} and U_2^{m+2} denote the functions which choose the first and second element from ordered pairs in $V_{m+2}(S)$, we write $c \in P$, $U_1^{m+2}(c) = a$, $U_2^{m+2}(c) = b$, instead of $P(a, b)$.] Let $b_1, \ldots, b_n$ be the constants in $\phi(x)$; $\phi(x)$ can now be written as $\phi(x, b_1, \ldots, b_n)$. Choose m so that $A, b_1, \ldots, b_n$ all belong to $*V_m(S)$. Then an unabbreviated form of the sentence

$$(\forall y_1 \in V_m(S)) \cdots (\forall y_n \in V_m(S))(\forall z \in V_m(S))$$
$$(\exists w \in V_{m+1}(S))(\forall x \in V_m(S))[x \in w \leftrightarrow x \in z \wedge \phi(x, y_1, \ldots, y_n)]$$

holds in $V(S)$ and therefore in $V(*S)$. Thus $\{x \in A : \phi(x)\} \in *V_{m+1}(S)$.

We should note again that all of the properties established for $*R$ in Section A are true for $*R$ in $V(*S)$. Note also that if $B = \{1, 2, 3\}$, then $*B = \{1, 2, 3\}$ since "$(\forall x \in B)[x = 1 \vee x = 2 \vee x = 3]$" is true for $V(S)$ and $V(*S)$. If $C = \{N, R\}$, then $*C = \{*N, *R\}$, but $*N \neq N$ and $*R \neq R$. In general we have the following result.

Theorem 2.7 *If B is a finite set $\{b_1, \ldots, b_n\}$ in $V(S)$, then $*B = \{*b_1, \ldots, *b_n\}$. If B is an infinite set in $V(S)$, then there is a $b \in *B$ such that for each $a \in B$, $*a \neq b$.*

Proof Let D be a hyperfinite set such that $D \subset *B$ and $*a \in D$ for each $a \in B$. Since $(\forall x \in \mathscr{P}_F(B))(\exists y \in B)[y \notin x]$ is true for $V(*S)$, there is a $b \in *B$ such that $b \notin D$, whence $b \neq *a$ for any $a \in B$.

Among the basic tools of Robinson's nonstandard analysis, are the following two results called permanence principles.

Theorem 2.8 *Let A be an internal set in $V(*S)$.*

(i) *If $A \supset N$, then there is an $\eta \in *N - N$ such that $n \in A$ for all $n \leq \eta$ in $*N$.*

(ii) *If $A \supset *N - N$, then there is a $k \in N$ such that $n \in A$ for all $n \geq k$ in $*N$.*

(iii) *If A contains the monad of r, $m(r)$, for some $r \in R$, then there is a $\delta > 0$ in R such that $\rho \in A$ for all $\rho \in *R$ with $|\rho - r| < \delta$.* (This is called Cauchy's principle in [27].)

Proof Assume $A \supset N$. Let $B = \{k \in *N : \forall n \leq k, n \in A\}$. Since B is internal by 2.6 and $B \supset N$, there is an $\eta \in B \cap (*N - N)$. The rest of the proof is left to the reader.

Theorem 2.9 *Let $\{a_n : n \in *N\}$ be an internal sequence in $V(*S)$.*

(i) *If $a_n \simeq 0$ for all $n \in N$, then for some $\gamma \in *N - N$ we have $a_n \simeq 0$ for all $n \leq \gamma$ in $*N$.*

(ii) *If a_n is infinite for all $n \in N$, then for some $\gamma \in *N - N$ it follows that a_n is infinite for all $n \leq \gamma$ in $*N$.*

Proof Let $B = \{n \in N : |a_n| \leq 1/n\}$. The set B is internal, so if $B \supset N$, then B contains an initial segment $\{1, \ldots, \gamma\}$ of $*N$ with $\gamma \in *N - N$. Part (ii) follows from the fact that $1/a_n \simeq 0$ if and only if a_n is infinite.

As an application of permanence principles, we consider a function $f: R \to R$ and an $r_0 \in R$ such that for each $\rho \simeq r_0$ we have $*f(\rho) \simeq f(r_0)$. Then given $\varepsilon > 0$ in R, the set $\{\rho \in *R : |*f(\rho) - f(r_0)| < \varepsilon\}$ is internal and contains the monad of r_0. Thus there is a $\delta > 0$ in R such that if $|\rho - r_0| < \delta$, then $|*f(\rho) - f(r_0)| < \varepsilon$. Other limit results from Section A can be obtained with equal ease using 2.8 and 2.9.

A final principle that we will often use is the fact that a standard sentence ψ that is true for $V(*S)$ is true for $V(S)$, since if $\neg \psi$ were true for $V(S)$, $\neg \psi$ would be true for $V(*S)$. For example, if $\{a_n : n \in N\}$ is a standard sequence in R and for some $\omega \in *N - N$, $a_\omega \simeq 0$, then 0 is a cluster point of $\{a_n\}$. This follows from the fact that if $m \in N$ and $\varepsilon > 0$ in R are given, then the sentence

$$(\exists n \in N)[n \geq m \wedge |a_n| < \varepsilon]$$

is true for $V(*S)$ and therefore for $V(S)$.

Given a topological space $(X, \mathcal{T})$ and an enlargement containing X, the monad $m(x)$ of a point $x \in X$ is the set $\bigcap_{0 \in \mathcal{T}, x \in 0} *0$. A set $U \subset X$ is open if $m(x) \subset *U$ for each $x \in U$. As in Theorem A. 4.3, a set $K \subset X$ is compact iff for each $y \in *K$ there is an $x \in K$ with $y \in m(x)$ (Robinson [24, Theorem 4.1.15]). For a product space $Y = \prod_{\alpha \in \mathcal{A}} X_\alpha$, the monad of a point $x = \{x_\alpha\} \in Y$ is

$$\prod_{\alpha \in \mathcal{A}} m(x_\alpha) \times \prod_{\alpha \in *\mathcal{A} - \mathcal{A}} X_\alpha.$$

If each X_α is compact and $y = \{y_\alpha\} \in \prod_{\alpha \in *\mathcal{A}} X_\alpha$, then there is an $x_\alpha \in X_\alpha$ for each standard $\alpha \in \mathcal{A}$ such that $y_\alpha \in m(x_\alpha)$. Let $x = \{x_\alpha\}$; then $y \in m(x)$. Thus we have Robinson's proof from [24] of the well-known fact that the product of compact spaces is compact.

III. A Nonstandard Representation of Measurable Spaces and L^∞

A. Representations of Measurable Spaces

In Section III we will apply the results of Section II to study standard measure spaces and L^∞ for those spaces. The discussion is patterned after that in [14], where further details can be found. The results of Section III will not be needed in later sections.

Let X be a fixed infinite set and $\mathcal{M}$ an infinite σ-algebra of subsets of X. We let $\mathcal{P}$ denote the collection of all finite $\mathcal{M}$-measurable partitions of X; i.e., $P \in \mathcal{P}$ if $P = \{A_1, \ldots, A_n\} \subset \mathcal{M}$, $X = \bigcup_{i=1}^n A_i$ and $A_i \cap A_j = \emptyset$ for $i \neq j$. We assume that $\emptyset \notin P$ for any $P \in \mathcal{P}$.

Let $V(S)$ be a superstructure containing X and the real numbers R, and let $V(*S)$ be a fixed enlargement of $V(S)$. Let $\mathscr{H}$ be a hyperfinite subset of $*\mathscr{P}$ such that for each $P \in \mathscr{P}$, $*P \in \mathscr{H}$. Fix an internal refinement P_0 of all of the partitions in $\mathscr{H}$. That is, $P_0 = \{A_1, \ldots, A_\omega\} \subset *\mathscr{M}$, $*X = \bigcup_{i=1}^{\omega} A_i$, $A_i \cap A_j = \varnothing$ when $i \neq j$; moreover, if $P = \{B_1, \ldots, B_\gamma\} \in \mathscr{H}$, then for each $B_j \in P$ and $A_i \in P_0$, either $A_i \subset B_j$ or $A_i \cap B_j = \varnothing$.

We shall let I be the initial segment $\{1, 2, \ldots, \omega\}$ of $*N$ such that $P_0 = \{A_i : i \in I\}$. For each standard $B \in \mathscr{M}$, we let $I_B = \{i \in I : A_i \subset *B\}$. Then $*B = \bigcup_{i \in I_B} A_i$ since P_0 is a refinement of $\{*B, *X - *B\} \in \mathscr{H}$.

If for each $x \in X$, $\{x\} \in \mathscr{M}$, then $\{x\} \in P_0$ for each standard $x \in X$. It follows that the external cardinality of P_0, and therefore $*N$, can be arbitrarily large since the cardinality of X can be arbitrarily large.

Theorem 1 *Let f be a real-valued bounded function on X. Then f is $\mathscr{M}$-measurable if and only if for each $i \in I$,*

(1) $$\sup_{x \in A_i} *f(x) - \inf_{x \in A_i} *f(x) \simeq 0.$$

Proof The functions sup and inf have nonstandard extensions which behave, relative to internal sets and $*R$, as sup and inf behave relative to standard sets and R. If f is $\mathscr{M}$-measurable, then given $\varepsilon > 0$ in R there is a partition $P_\varepsilon \in \mathscr{P}$ such that for each $B_j \in P_\varepsilon$,

$$\sup_{x \in B_j} f(x) - \inf_{x \in B_j} f(x) < \varepsilon.$$

Since P_0 is a refinement of $*P_\varepsilon$ and ε is arbitrary, equation (1) holds. If f is not $\mathscr{M}$-measurable, then f cannot be approximated by simple functions. Therefore, for some $\varepsilon > 0$ in R we have for each $P \in \mathscr{P}$ a $B_j \in P$ with

$$\sup_{x \in B_j} f(x) - \inf_{x \in B_j} f(x) \geq \varepsilon.$$

It follows that for some $A_i \in P_0$,

$$\sup_{x \in A_i} *f(x) - \inf_{x \in A_i} *f(x) \geq \varepsilon.$$

Definition 2 Let MB denote the collection of all bounded $\mathscr{M}$-measurable real-valued functions on X. Let E denote the set of all internal mappings from the index set I into $*R$, and let $\cong$ be the external equivalence relation in E such that $x \cong y$ if and only if for each $i \in I$, $x(i) \simeq y(i)$. Let C_{p_0} be a fixed internal function on I such that for each $i \in I$, $C_{p_0}(i) \in A_i$. Let T denote the mapping from MB into E such that for each $f \in MB$ and $i \in I$, $T(f)(i) = *f(C_{p_0}(i))$.

The set E is internal Euclidean ω-space in $V(*S)$; using the mapping T, we treat each $f \in MB$ as a vector in E. Given $f \in MB$, the choice of $T(f) \in E$

is unique up to the relation $\cong$. Moreover, it is easy to see that if f and g are in MB and α and β are in R, then $T(\alpha f + \beta g) = \alpha T(f) + \beta T(g)$, $T(fg) = T(f)T(g)$, i.e., $T(fg)(i) = T(f)(i)T(g)(i)$ for each $i \in I$, and $T(f) \not\cong T(g)$ if $f \neq g$.

Recall that a function $\mu : \mathcal{M} \to R$ is finitely additive if for each pair B_1, $B_2 \in \mathcal{M}$ with $B_1 \cap B_2 = \emptyset$ we have $\mu(B_1 \cup B_2) = \mu(B_1) + \mu(B_2)$; of course, $\mu(\emptyset) = 0$.

Definition 3 Let $\Phi(X, \mathcal{M})$, or simply Φ, denote the collection of finitely additive real-valued functions μ on $\mathcal{M}$ with $\sup_{B \in \mathcal{M}} |\mu(B)| < +\infty$. For each $\mu \in \Phi$, let $U(\mu)$ be the internal vector in E such that $U(\mu)(i) = \mu(A_i)$ for each $i \in I$. If $e \in E$ and $^0\sum_{i \in I} |e(i)| < +\infty$, let $\phi(e)$ be the element of Φ defined by setting $\phi(e)(B) = {^0\sum_{i \in I_B}} e(i)$ for each $B \in \mathcal{M}$.

Note that $\sum$ is the internal summation operation on hyperfinite sets; it is the extension of $\sum$ on finite sets.

Proposition 4 *Fix μ and $v \in \Phi$ and α and β in R. Then $U(\alpha \mu + \beta v) = \alpha U(\mu) + \beta U(v)$ and $\phi(U(\mu))) = \mu$.*

Proof Let $b = \sup_{B \in \mathcal{M}} |\mu(B)|$. Then $\sum_{i \in I} |U(\mu)(i)| \leq 2b$ in $V(*S)$. Given $B \in \mathcal{M}$, we have

$$\phi(U(\mu))(B) \simeq \sum_{i \in I_B} U(\mu)(i) = \sum_{i \in I_B} {}^*\mu(A_i) = {}^*\mu(*B) = \mu(B).$$

Since $\phi(U(\mu))(B)$ is standard, it equals $\mu(B)$. The rest of the proof is clear.

We will show that for some $e \in E$, $U(\phi(e)) \not\cong e$. The relationship between the lattice operations on Φ and on E is easy to establish. For example, $\mu \vee v = \phi(U(\mu) \vee U(v))$, and the total variation $|\mu|(X)$ is $\sum_{i \in I}^{0} |U(\mu)(i)|$.

We next show that for each $\mu \in \Phi$ there is a null set that contains every standard null set. Recall that $B \in \mathcal{M}$ is a μ-positive set (or a μ-negative set) if for each $C \in \mathcal{M}$ with $C \subset B$ we have $\mu(C) \geq 0$ [or $\mu(C) \leq 0$]. A μ-null set is a set that is both μ-positive and μ-negative.

Theorem 5 *Fix $\mu \in \Phi$. Let $A_+ = \cup \{A_i \in P_0 : {}^*\mu(A_i) > 0\}$, $A_- = \cup \{A_i \in P_0 : {}^*\mu(A_i) < 0\}$, and $A_0 = \cup \{A_i \in P_0 : {}^*\mu(A_i) = 0\}$. Then ${}^*\mu(A_0) = 0$ and for each standard μ-null set B we have ${}^*B \subset A_0$. If μ is countably additive and $\{B_+, B_-\}$ is a standard Hahn decomposition of X with respect to μ, then $A_+ \subset {}^*B_+$, $A_- \subset {}^*B_-$, and for each internal ${}^*\mathcal{M}$-measurable $C \subset A_0$, ${}^*\mu(C) = 0$.*

Proof If B is a standard μ-null set and $i \in I_B$, then ${}^*\mu(A_i) = 0$, whence $A_i \in A_0$. It follows that ${}^*B = \bigcup_{i \in I_B} A_i \subset A_0$. Clearly, ${}^*\mu(A_0) = \sum_{A_i \subset A_0} {}^*\mu(A_i) = 0$. If $\{B_+, B_-\}$ is a Hahn decomposition of X for μ, and $A_i \subset A_0$, then since either $A_i \subset {}^*B_+$ or $A_i \subset {}^*B_-$, A_i is a ${}^*\mu$-null set. The rest of the proof is clear.

If $X = R$ and λ is Lebesgue measure on the Lebesgue measurable sets $\mathcal{M}$ in R, then for each $r \in R$, $\{r\} \in P_0$ and $r \in A_0$. On the other hand, if $X = N$ and $\mathcal{M}$ is the power set of N, then $\{n\} \in P_0$ for each $n \in N$. Let γ be the first $n \in {}^*N$ such that $\{n\} \notin P_0$, and let $A_i = \{i\}$ for all $i < \gamma$. Let $e \in E$ be given by setting $e_i = 2^{-i}$ for $1 \leq i \leq \gamma - 2$, $e_{\gamma-1} = -2$, and $e_i = 0$ for $i \geq \gamma$. For each $n \in N$, $\phi(e)(\{n\}) = 2^{-n}$ while $\phi(e)(N) = -1$. Therefore, $\phi(e)$ has no standard Hahn decomposition. Moreover, $^*\phi(e)(\{\gamma - 1\}) > 0$, so $U(\phi(e)) \not\simeq e$. Let $K = \bigcup_{i<\gamma} A_i \subset {}^*N$. If μ is countably additive in Φ, then $^*\mu({}^*N - K) \simeq 0$, while if μ is purely finitely additive, then $^*\mu(K) = 0$ (see Robinson [23] or [14, pp. 69–70]).

In the general case, it is fairly easy to establish the following relationship between integration and the inner product $x \cdot y \to \sum_{i \in I} x_i y_i$ in E.

Theorem 6 *Fix* $f \in MB$ *and* $\mu \in \Phi$. *For each* $B \in M$,

$$\int_B f \, d\mu = \sum_{i \in I_B}^{0} {}^*f(C_{P_0}(i))^*\mu(A_i).$$

In particular,

$$\int_X f \, d\mu = {}^0(T(f) \cdot U(\mu)).$$

Proof See Loeb [14, Theorem 3.2].

We have shown that for each $\mu \in \Phi$ and $B \in \mathcal{M}$, $\mu(B) = {}^0\sum_{i \in I_B} U(\mu)(i) = {}^0\sum_{i \in I_B} {}^*\mu(A_i)$. If we concentrate the weight $^*\mu(A_i)$ at the point $C_{P_0}(i)$ and let $\mu(B) = {}^0\sum_{C_{P_0}(i) \in {}^*B} {}^*\mu(A_i)$ for any set $B \subset X$, then we have a finitely additive extension of μ to $\mathcal{P}(X)$. Such a hyperfinite set of weighted points is called a sample. A sample was first used by Bernstein and Wattenberg [5] to represent Lebesgue measure on $[0, 1]$, and later by Parikh and Parness [21] to study conditional probabilities and Henson [7] to study the integration of unbounded functions.

Given a finite measure μ on $(X, \mathcal{M})$, one can construct special partitions of the form P_0 as in [15] and [4]. For example, let $\mathcal{F}$ be a hyperfinite set of internal $^*\mu$-integrable functions containing the extension of each standard μ-integrable function. Let $g = \sum_{f \in \mathcal{F}} |f|$. Then g is $^*\mu$-integrable. Choose $\eta \in {}^*N - N$ so that if $\tilde{A} = \{x \in {}^*X : |g(x)| \geq \eta\}$, then $^*\mu(\tilde{A}) \simeq 0$ and $\int_{*X} g \, d^*\mu \simeq \int_{*X - \tilde{A}} g \, d^*\mu$. We may assume that P_0 is a refinement of the partition $\{\tilde{A}, *X - \tilde{A}\}$ and that for each $A_i \subset {}^*X - \tilde{A}$ and each $f \in \mathcal{F}$, $\sup_{A_i} f(x) - \inf_{A_i} f(x) \simeq 0$. It follows that for each $f \in \mathcal{F}$, in particular for the extension of each standard integrable function,

$$\int_{*X} f \, d^*\mu \simeq \int_{*X - \tilde{A}} f \, d^*\mu \simeq \sum_{A_i \subset {}^*X - \tilde{A}} {}^*f(C_{P_0}(i))^*\mu(A_i).$$

If μ is a nonatomic probability measure, we can also choose P_0 so that for each $B \in \mathcal{M}$, $\mu(B) \simeq |I_B|/|I|$, where $|I_B|$ denotes the internal coordinality of I_B.

B. A Representation of L^∞ and Its Dual

Let $\mathcal{N}$ be a proper subset of $\mathcal{M}$ such that countable unions of sets in $\mathcal{N}$ are in $\mathcal{N}$ and subsets of sets in $\mathcal{N}$ are in $\mathcal{N}$. For example, $\mathcal{N}$ may be the null sets for a measure on $(X, \mathcal{M})$. For each $\mathcal{M}$-measurable function f, let

$$\|f\|_\infty = \inf\{\alpha \in R : \{x \in X : |f(x)| > \alpha\} \in \mathcal{N}\},$$

and let M_0 be the set of $\mathcal{M}$-measurable f's with $\|f\|_\infty < +\infty$. As usual, we call f and $g \in M_0$ equivalent if $\|f - g\|_\infty = 0$ and we let L^∞ denote the equivalence classes in M_0.

Let $I_0 = \{i \in I : A_i \in {}^*\mathcal{N}\}$, and for each $f \in M_0$, let $T_0(f)(i) = T(f)(i)$ when $i \notin I_0$, while $T_0(f)(i) = 0$ if $i \in I_0$. It is easy to see that for an $\mathcal{M}$-measurable B, $B \in \mathcal{N}$ if and only if $I_B \subset I_0$. If f and g are in M_0, then $T_0(f + g) = T_0(f) + T_0(g)$, $T_0(\alpha f) = \alpha T_0(f)$ for each $\alpha \in R$, $T_0(fg) = T_0(f)T_0(g)$, and $\|f\|_\infty \simeq \max_{i \in I}|T_0(f)(i)|$. Moreover, if $\|f - g\|_\infty = 0$, then $T_0(f) = T_0(g)$.

Let $(L^\infty)^d$ denote the dual space of L^∞. For each $F \in (L^\infty)^d$, let $V(F)$ be the element of E such that $V(F)(i) = {}^*F(\chi_{A_i})$ for $i \in I$. Let μ_F denote the element of Φ such that $\mu_F(B) = F(\chi_B)$ for each $B \in \mathcal{M}$. Then $\mu_F = \phi(V(F))$ and $U(\mu_F) = V(F)$. Let Φ_0 denote the normed vector space $\{\mu \in \Phi : \mu(B) = 0 \; \forall B \in \mathcal{N}\}$ with norm $\|\mu\| = |\mu|(X)$. It is easy to establish [14, Theorem 4.4] the well-known fact ([6, 28]) that the mapping $F \to \mu_F$ is an isometric isomorphism from $(L^\infty)^d$ onto Φ_0. For each $F \in (L^\infty)^d$ and $f \in L^\infty$,

$$F(f) = \int_X f \, d\mu_F \simeq V(F) \cdot T_0(f).$$

Recall that an element $F \in (L^\infty)^d$ is multiplicative if for each $f, g \in L^\infty$, $F(f \cdot g) = F(f) \cdot F(g)$; in this case, μ_F takes only the values 0 and 1. Let Φ_1 denote the set of nonzero multiplicative functionals on L^∞. Given $F \in \Phi_1$, there is an $i_F \in I - I_0$ such that $V(F)(i_F) = 1$ and $V(F)(i) = 0$ for $i \neq i_F$. Conversely, if $\delta^i \in E$ is zero at all $j \neq i \notin I_0$ and $\delta^i(i) = 1$, then $\phi(\delta^i) = \mu_F$ for some $F \in \Phi_1$, but $V(F) = U(\mu_F)$ need not equal δ^i. If we call elements i and j of $I - I_0$ equivalent when $\phi(\delta^i) = \phi(\delta^j)$, that is, when ${}^*f(C_{P_0}(i)) \simeq {}^*f(C_{P_0}(j))$ for all $f \in L^\infty$, then the equivalence classes in $I - I_0$ form the Stone space for L^∞.

To see that the Stone space for L^∞ is compact, we consider a nonstandard $F \in {}^*\Phi_1$ and note that for some $i \in I - I_0$, $F(\chi_{A_i}) = 1$ while $F(\chi_{A_j}) = 0$ for all $j \neq i \in I$. It follows that F is in the monad of $\phi(\delta^i)$ with respect to the topology generated by L^∞. For more details, see [14] and the subsequent developments in [27].

IV. Conversion from Nonstandard to Standard Measure Spaces

A. Measure Spaces and Integration

Let $V(*S)$ be a fixed enlargement of a superstructure $V(S)$. We make the assumption that if $\{A_n : n \in N\}$ is an ordinary sequence of internal objects contained in $V_m(*S)$ for some $m \in N$, then there is an internal sequence $\{B_n : n \in *N\}$ such that $A_n = B_n$ for all $n \in N$. Enlargements, such as ultrapowers, with this property are said to be $\aleph_1$-saturated or denumerably comprehensive. The following discussion is patterned after [16].

Let X be an internal set in $V(*S)$ and let $\mathscr{A}$ be an internal algebra of subsets of X. Then $\mathscr{A}$ is an algebra in the usual sense. That is, if A and B are in $\mathscr{A}$, so are $A \cup B$ and $X - A$; so also is any hyperfinite union of elements of $\mathscr{A}$. For ordinary countable unions, we have the following result.

Proposition 1 Fix $A_n \in \mathscr{A}$ for each $n \in N$ and assume that $A_0 \subset \bigcup_{n=1, n \in N}^{\infty} A_n$. Then for some $m \in N$, $A_0 \subset \bigcup_{n=1}^{m} A_n$.

Proof Let $\{A_n : n \in *N\}$ be an internal extension of the sequence $\{A_n : n \in N\}$. Then the internal set $\{m \in *N : A_0 \subset \bigcup_{n=1}^{m} A_n\}$ has a first element which must be finite.

We now let v be an internal mapping of $\mathscr{A}$ into $*R^+ = \{r \in *R : r \geqslant 0\}$ such that v is finitely additive, i.e., $v(\varnothing) = 0$ and $v(A \cup B) = v(A) + v(B)$ when A and B are disjoint elements of $\mathscr{A}$. Let v_0 be the extended real-valued finitely additive measure defined on the algebra $\mathscr{A}$ by setting $v_0(A) = {}^0(v(A))$ for each $A \in \mathscr{A}$. (Recall that for a positive infinite $\rho \in *R$, ${}^0\rho = +\infty$.) Let $\sigma(\mathscr{A})$ be the ordinary σ-algebra of subsets (both internal and external) of X such that $\sigma(\mathscr{A}) \supset \mathscr{A}$ and for any other σ-algebra $\mathscr{B} \supset \mathscr{A}$, $\mathscr{B} \supset \sigma(\mathscr{A})$. This means that ordinary countable unions (i.e., with respect to N) of sets in $\sigma(\mathscr{A})$ are again in $\sigma(\mathscr{A})$.

Theorem 2 *The finitely additive measure v_0 on $\mathscr{A}$ has a unique countably additive extension, also denoted by v_0, defined on $\sigma(\mathscr{A})$. For each $B \in \sigma(\mathscr{A})$, $v_0(B) = \inf_{A \in \mathscr{A}} v_0(A)$, and if $v_0(B)$ is finite, then $v_0(B) = \sup_{A \in \mathscr{A}, A \subset B} v_0(A)$ and there is an $A \in \mathscr{A}$ with $v_0((B - A) \cup (A - B)) = 0$.*

Proof By Proposition 1, a countable, infinite, disjoint union of non-empty sets from $\mathscr{A}$ is not an element of $\mathscr{A}$. Thus v_0 is actually countably additive on $\mathscr{A}$. By the Carathéodory extension theorem [25, pp. 253–260], v_0 has a countably additive extension, v_0, defined on $\sigma(\mathscr{A})$. In general, the extension is unique when $v_0(X) < +\infty$; Henson [9] has shown that in our case it is also unique when $v_0(X) = +\infty$. His proof uses the fact that for each $B \in \sigma(\mathscr{A})$ there is a σ-algebra containing B generated by some countable algebra $\tilde{\mathscr{A}} \subset \mathscr{A}$; it shows that either for some $A \in \mathscr{A}$, $A \subset B$ and $v_0(A) = +\infty$, or for some sequence $\{A_n : n \in N\} \subset \mathscr{A}$ with $v_0(A_n) < +\infty$ for each $n \in N$ we have $B \subset \bigcup_{n=1}^{\infty} A_n$. In any case, given $B \in \sigma(\mathscr{A})$ with $v_0(B) < +\infty$ and given $\varepsilon > 0$ in R, there is a sequence $\{A_n : n \in N\} \subset \mathscr{A}$ such that $A_n \subset A_{n+1}$ for each $n \in N$, $C = \bigcup_{n=0}^{\infty} A_n \supset B$, and $v_0(C) < v_0(B) + \varepsilon$. Let $\{A_n : n \in *N\}$ be an internal extension of this sequence. There is a $\gamma \in *N - N$ such that for all $n \in *N$ with $1 \leqslant n \leqslant \gamma$ we have $A_{n-1} \subset A_n$, $A_n \in \mathscr{A}$, and $v_0(A_n) < v_0(B) + \varepsilon$. Now $B \subset C \subset A_\gamma$, and ε is arbitrary, so $v_0(B) = \inf_{A \in \mathscr{A}, B \subset A} v_0(A)$. It follows also, by working with $A_\gamma - B$, that $v_0(B) = \sup_{A \in \mathscr{A}, A \subset B} v_0(A)$. Moreover, for each $n \in N$ there are sets $\tilde{A}_n$ and A_n in $\mathscr{A}$ with

$$\tilde{A}_{n-1} \subset \tilde{A}_n \subset B \subset A_n \subset A_{n-1}$$

and $v_0(A_n - \tilde{A}_n) \leqslant 1/n$ for $n \geqslant 1$. Thus, there is a set $A_\omega \in \mathscr{A}$ such that for each $n \in N$, $\tilde{A}_n \subset A_\omega \subset A_n$, whence $v_0((B - A_\omega) \cup (A_\omega - B)) = 0$.

Let, for example, $\mathscr{A}$ be the extension $*\mathscr{L}$ of the Lebesgue measurable sets $\mathscr{L}$ in R, and let v be the extension $*\lambda$ of Lebesgue measure on $\mathscr{L}$. Then $C = \bigcup_{n=1, n \in N}^{\infty} *[-n, n]$ is $\sigma(\mathscr{A})$-measurable and $v_0(C) = +\infty$, but $C \neq *R$. The monad of zero $m(0) = \bigcap_{n=1, n \in N}^{\infty} *[-1/n, 1/n]$ is also $\sigma(\mathscr{A})$-measurable, and $v_0(m(0)) = 0$. Since R is contained in a hyperfinite set in $*R$, R is contained in a v_0-null set.

We next show that internal, $\mathscr{A}$-measurable functions can be converted to extended real-valued $\sigma(\mathscr{A})$-measurable functions on X. Given $f : X \to *R$, we let ^0f take the value $^0(f(x))$ at each $x \in X$. Such an f is $\mathscr{A}$-measurable if it is internal and for each $a \in *R$, $\{x \in X : f(x) < a\} \in \mathscr{A}$ and $\{x \in X : f(x) \leqslant a\} \in \mathscr{A}$.

Theorem 3 *If $f : X \to *R$ is $\mathscr{A}$-measurable, then $^0f : X \to R \cup \{+\infty, -\infty\}$ is $\sigma(\mathscr{A})$-measurable.*

Proof For each standard $\alpha \in R$,

$$\{x \in X : {}^0f(x) < \alpha\} = \bigcup_{n=1, n \in N}^{\infty} \{x \in X : f(x) < \alpha - (1/n)\} \in \sigma(\mathscr{A}).$$

This result has obvious generalizations which are left to the reader (see [16] and [1]).

For simplicity, we will assume from this point on that $v_0(X) < +\infty$. We say that $f : X \to *R$ is finite valued if $f(x)$ is finite for all $x \in X$. If f is both internal and finite valued, then for some $n \in N$, $|f(x)| < n$ for all $x \in X$ by Theorem II.B.2.8(ii).

Theorem 4 *If f is a finite-valued, internal, $\mathcal{A}$-measurable function on X, then for each $A \in \mathcal{A}$,*

$$\int_A f\, dv \simeq \int_A {}^0\!f\, dv_0.$$

Proof Since ${}^0(f \cdot \chi_A) = {}^0\!f \cdot \chi_A$, we need only consider the case $A = X$. We may also assume that $f(x) \geq 1$ for all $x \in X$, since if $f + n \geq 1$ for $n \in \mathbb{N}$ and $\int f + n\, dv \simeq \int {}^0\!f + n\, dv_0$, then since $\int n\, dv \simeq \int n\, dv_0$, $\int f\, dv \simeq \int {}^0\!f\, dv_0$. Let $D = \{r \in R : v_0({}^0\!f^{-1}[r]) > 0\}$; D is finite or countably infinite in R. Let $M = v_0(X) + 1$ and choose $\varepsilon > 0$ in R. There is a finite increasing sequence $0 = y_0 < y_1 \cdots < y_m$ in R with $y_m > \sup_{x \in X} {}^0\!f(x)$ such that $y_i \notin D$ and $y_i - y_{i-1} < \varepsilon/3M$ for $1 \leq i \leq m$. Let $f^{-1}[a, b)$, ${}^0\!f^{-1}[a, b)$, etc., denote sets such as $\{x \in X : a \leq f(x) < b\}$, $\{x \in X : a \leq {}^0\!f(x) < b\}$, etc. Let

$$S_v = \sum_{i=1}^m y_{i-1} v(f^{-1}[y_{i-1}, y_i)), \qquad \bar{S}_v = \sum_{i=1}^m y_i v(f^{-1}[y_{i-1}, y_i)),$$

$$S_{v_0} = \sum_{i=1}^m y_{i-1} v_0({}^0\!f^{-1}[y_{i-1}, y_i)), \qquad \bar{S}_{v_0} = \sum_{i=1}^m y_i v_0({}^0\!f^{-1}[y_{i-1}, y_i)).$$

Then

$$S_v \leq \int_X f\, dv \leq \bar{S}_v, \qquad S_{v_0} \leq \int_X {}^0\!f\, dv_0 \leq \bar{S}_{v_0},$$

$$\bar{S}_v - S_v \leq \frac{\varepsilon}{3M} \sum_{i=1}^m v(f^{-1}[y_{i-1}, y_i)) < \frac{\varepsilon}{3},$$

and $\bar{S}_{v_0} - S_{v_0} < \varepsilon/3$. Given $i \in \mathbb{N}$ with $1 \leq i \leq m$,

$${}^0\!f^{-1}(y_{i-1}, y_i) \subseteq f^{-1}(y_{i-1}, y_i) \subseteq f^{-1}[y_{i-1}, y_i] \subseteq {}^0\!f^{-1}[y_{i-1}, y_i].$$

Therefore,

$$v_0({}^0\!f^{-1}[y_{i-1}, y_i]) = v_0({}^0\!f^{-1}(y_{i-1}, y_i)) = v_0({}^0\!f^{-1}(y_{i-1}, y_i))$$
$$\leq v_0(f^{-1}(y_{i-1}, y_i)) \simeq v(f^{-1}(y_{i-1}, y_i)) \leq v(f^{-1}[y_{i-1}, y_i))$$
$$\leq v(f^{-1}[y_{i-1}, y_i]) \simeq v_0(f^{-1}[y_{i-1}, y_i])$$
$$\leq v_0({}^0\!f^{-1}[y_{i-1}, y_i]).$$

It follows that $S_v \simeq S_{v_0}$, whence $|\int_X f\, dv - \int_X {}^0\!f\, dv_0| < \varepsilon$. Since ε is arbitrary, the theorem is proved.

In [16], the author considered the example of a point $x_1 \in X$ with $v(\{x_1\}) = \delta \not\simeq 0$ and the function $f = (1/\delta)\chi_{\{x_1\}}$. For this case, $\int f\, dv = 1$ but $\int {}^0\!f\, dv_0 = 0$. By the definition of the integral and Theorem 4, if $A \in \mathcal{A}$ and f is $\mathcal{A}$-measurable, then

$$\int_A {}^0|f|\, dv_0 \equiv \lim_{n \in \mathbb{N}, n \to \infty} \int_A {}^0|f| \wedge n\, dv_0 = \lim_{n \in \mathbb{N}, n \to \infty} {}^0\!\int_A |f| \wedge n\, dv.$$

Anderson [1] showed that this limit equals $^0\!\int_A |f|\, dv$ for all $A \in \mathscr{A}$ if and only if $^0\!\int_A |f|\, dv$ is finite for $A = X$ and infinitesimal when $v(A) \simeq 0$. Following precedent, he called this condition S-integrability. We shall modify this definition as follows:

Definition 5 An $\mathscr{A}$-measurable $f: X \to {}^*R$ is S-integrable if for each $\eta \in {}^*N - N$, $\int_{\{|f| \geq \eta\}} |f|\, dv \simeq 0$.

Now the following result is a clear consequence of the definition of the integral and Theorem 4.

Theorem 6 An $\mathscr{A}$-measurable $f: X \to {}^*R$ is S-integrable if and only if $^0\!\int_X |f|\, dv$ is finite and equals

$$\int_X {}^0|f|\, dv_0 \equiv \lim_{n \in N, n \to \infty} \int_X {}^0|f| \wedge n\, dv_0 = \lim_{n \in N, n \to \infty} {}^0\!\int_X |f| \wedge n\, dv.$$

If f is S-integrable, then for each $A \in \mathscr{A}$, $^0\!\int_A f\, dv = \int_A {}^0f\, dv_0$. In particular, if $v(A) \simeq 0$, then (since $v_0(A) = 0$) $\int_A f\, dv \simeq 0$.

If $v_0(X) = +\infty$, we add the additional requirements for S-integrability that $^0\!\int |f|\, dv < +\infty$ and, as in [1], that $\int_{\{|f| \leq 1/\eta\}} |f|\, dv \simeq 0$ for each $\eta \in {}^*N - N$. It is helpful here to work with a sequence $a_n \downarrow 0$ such that $v_0({}^0f^{-1}(a_n)) = 0$ for each $n \in N$.

Recall that a standard sequence $\{f_m\}$ of measurable, extended real-valued functions on a finite measure space $(Y, \mathscr{M}, \mu)$ is called uniformly integrable if for each $\varepsilon > 0$ in R there is a $k \in N$ such that $\int_{\{|f_m| \geq k\}} |f_m|\, d\mu < \varepsilon$ for all $m \in N$. The following theorem is clear (and noted independently by Anderson [2]).

Theorem 7 A sequence $\{f_m : m \in N\}$ is uniformly integrable if and only if for the nonstandard extension of the sequence, $\{f_m : m \in {}^*N\}$, f_m is S-integrable for all $m \in {}^*N$.

Corollary 8 If $\{f_m : m \in N\}$ is uniformly integrable on a finite measure space $(Y, \mathscr{M}, \mu)$ and $f_m \to g$ in measure, then $\int f_m\, d\mu \to \int g\, d\mu < +\infty$.

Proof Given any $\eta \in {}^*N - N$, there is a positive $\varepsilon \simeq 0$ such that ${}^*\mu(\{|f_\eta - g| \geq \varepsilon\}) \leq \varepsilon$. Since for $\eta \in {}^*N - N$

$$\int g\, d\mu = \lim_{m,n \in N,\, m,n \to \infty} \int (m \wedge g \vee -n)\, d\mu$$

$$= \lim_{m,n \in N,\, m,n \to \infty} {}^0\!\int (m \wedge f_\eta \vee -n)\, d{}^*\mu \simeq \int f_\eta\, d{}^*\mu,$$

$$\lim_{n \to \infty} \int f_n\, d\mu = \int g\, d\mu < +\infty.$$

Results similar to Theorem 7 and Corollary 8 are, of course, valid for the case that $v_0(X) = +\infty$.

Theorem 9 *Let (Y, d) be a metric space in the superstructure $V(S)$, and let f be a $\sigma(\mathscr{A})$-measurable mapping of X into Y. Then there is an internal, $\mathscr{A}$-measurable $g: X \to {}^*Y$ such that ${}^*d(g(x), f(x)) \simeq 0$, i.e., ${}^0g(x) = f(x)$, v_0 - a.e.*

Proof By the measurability of f, we mean that there is a sequence of $\sigma(\mathscr{A})$-measurable simple functions f_n such that $f_n(x) \to f(x)$ for all $x \in X$. By Theorem 2, for each $n \in N$ there is an internal, $\mathscr{A}$-measurable h_n equal to f_n, v_0-a.e. Given $\varepsilon > 0$ in R, there is (Egoroff's theorem) an $A \in \mathscr{A}$ with $v(X - A) < \varepsilon$ such that each f_n is internal on A and $f_n \to f$ uniformly on A. Extend the sequence $\{f_n|A : n \in N\}$ to an internal sequence of $\mathscr{A}$-measurable functions on A. For sufficiently small $\eta \in {}^*N - N$, ${}^0f_\eta = f|A$. We can thus find internal hyperfinite sequences $\{A_n : 1 \leq n \leq \gamma\}$ and $\{g_n : 1 \leq n \leq \gamma\}$ such that for $n \leq m \leq \gamma$, $A_n \in \mathscr{A}$, $A_n \cap A_m = \varnothing$ if $n \neq m$, $v(X - \bigcup_{n=1}^m A_n) \simeq 0$ for $m \in {}^*N - N$ each g_n is $\mathscr{A}$-measurable and has domain A_n, and ${}^0g_n = f|A_n$ if $n \in N$. Choose $y_0 \in Y$ and let $g(x) = g_n(x)$ if $x \in A_n$ and $g(x) = y_0$ if $x \in X - \bigcup_{n=1}^\gamma A$. Then ${}^0g = f$, v_0-a.e.

In [16], the author included this corollary of Theorem 2 for the case $Y = R$ at the suggestion of L. C. Moore, Jr. The replacement of f with an $\mathscr{A}$-measurable g is called a lifting by Anderson [1]. The existence of a lifting for an extended real-valued function is established in [16], and the result is still valid when $v_0(X) = +\infty$ if $v_0(\{|f| \geq 1/n\}) < +\infty$ for each $n \in N$. If $v_0(X) < +\infty$ and f is v_0-integrable on X, then given a lifting g of f, it follows that for sufficiently small $\eta \in {}^*N - N$, $-\eta \vee g \wedge \eta$ is an S-integrable lifting of f. [For the analogous development when $v_0(X) = +\infty$, consider a sequence $a_n \downarrow 0$ with $v_0(f^{-1}(a_n)) = 0$ for each $n \in N$.]

The existence of S-integrable liftings is established in [1] as part of an L^p-theory developed there. Liftings are obtained for functions with range in a Hausdorff space with countable base in [2]. In Anderson's [1] representation of the Itô Integral, he noted and used the fact that if $\{A_1, \ldots, A_k\}$ is a hyperfinite partition of X and $\mathscr{A}$ consists of all internal unions of the A_i's, then a lifting of a $\sigma(\mathscr{A})$-measurable f must be constant on the A_i's since it is $\mathscr{A}$-measurable.

For hyperfinite probability spaces we will be using the following form of the theory we have developed so far.

Theorem 10 *Let $Y = \{y_i : 1 \leq i \leq \omega\}$ and $\{\beta_i \in {}^*R : 1 \leq i \leq \omega\}$ be hyperfinite sets of the same internal cardinality in an $\aleph_1$-saturated enlargement. Assume that $\beta_i \geq 0$ for each $i \leq \omega$ and ${}^0\sum_{i=1}^\omega \beta_i < +\infty$. Let $f: Y \to {}^*[-n, n]$*

for some $n \in N$ be internal and let $v(A) = \sum_{y_i \in A} \beta_i$ for each internal $A \subset Y$. Then there is a unique extension v_0 of ^0v to the smallest (external) σ-algebra containing all internal subsets of Y, and

$$\int_Y {}^0f(y)\,dv_0(y) \simeq \sum_{i=1}^{\omega} f(y_i)\beta_i.$$

B. The Standard Part Map

Let $(X, \mathcal{T})$ be a compact Hausdorff space, and let $C(X)$ denote the continuous real-valued functions on X. Let $\mathcal{B}(X)$ denote the Baire subsets of X, i.e., the smallest σ-algebra for which each $f \in C(X)$ is measurable. Fix an enlargement $V(*S)$ of a superstructure containing X and R. Recall that the monad $m(x)$ of an $x \in X$ is the set $\bigcap_{0 \in \mathcal{T}, x \in 0} *0$; for each $y \in *X$ there is a (unique) $x \in X$ with $y \in m(x)$ since X is compact. We write $x = {}^0y$ or $x = \text{st}(y)$ when $y \in m(x)$. We may assume that $V(*S)$ has the property that for each internal set $A \subset *X$, $\text{st}(A) = \{\text{st}(y) : y \in A\}$ is compact in X. This is always the case if $(X, \mathcal{T})$ is metrizable (Robinson [24]) and follows with "saturation" in the general case (Luxemburg [20]). The following two results are taken from [19].

Theorem 1 *Let Z be an internal subset of $*X$. Let $\mathcal{A}$ denote the algebra of internal Baire sets in Z, and let $\sigma(\mathcal{A})$ be the smallest external σ-algebra containing $\mathcal{A}$. Let $K = \text{st}(Z)$ and $\mathcal{B}(K)$ be the Baire sets in K. Then the standard part map $\text{st}_Z : Z \to K$, where $\text{st}_Z(x) = {}^0x$ for each $x \in Z$, is measurable with respect to $\sigma(\mathcal{A})$ and $\mathcal{B}(K)$. Moreover, if v is an internal, nonnegative $\mathcal{A}$-measure on Z with ${}^0(v(Z)) < +\infty$, and v_0 is the corresponding σ-additive measure on $(Z, \sigma(\mathcal{A}))$ given by Theorem A.2, then letting $v_K(B) = v_0(\text{st}_Z^{-1}(B))$ for each $B \in \mathcal{B}(K)$, it follows that v_K is a Baire measure on K and for each $f \in C(X)$,*

$$(2) \quad \int_K f\,dv_K = \int_Z {}^0(*f)\,dv_0 \simeq \int_Z *f\,dv.$$

Proof Given $f \in C(K)$, extend f continuously to all of X. For each $\alpha \in R$,

$$\text{st}_Z^{-1}(\{x \in K : f(x) < \alpha\}) = \{z \in Z : {}^0(*f(z)) < \alpha\}$$

$$= \bigcup_{n=1, n \in N}^{\infty} \{z \in Z : *f(z) < \alpha - (1/n)\} \in \sigma(\mathcal{A}).$$

It follows that st_Z is measurable. If $f \in C(X)$ and $y \in Z$, then ${}^0(*f(y)) = f(\text{st}(y))$, whence equation (2) follows.

Corollary 2 *If v is an internal Baire measure on $*X$ and $K \in \mathcal{B}(X)$, then v_K is the standard part of v with respect to the weak* topology for Baire measures on X, i.e., the topology generated by $C(X)$.*

A special case of Theorem 1 was established by the author and widely communicated at the same time as the basic results in Section A. The space X was a product space, and only point evaluations were considered in the analog of Equation (2). This result eventually appeared as Theorem 4.6 of [17]. Later, Anderson [1] used the standard part map to construct Wiener measure on $C([0,1])$ and Lebesgue measure on $[0,1]$. The coincidence of the standard part map and ordinary weak* convergence methods as means of obtaining standard measures was then demonstrated by Loeb's Theorem 6.5 [17], Anderson's proof of Donsker's theorem for Brownian motion [1], and the following result established by Rashid for a compact metric space [22]. Here, $M(X)$ denotes the set of finite signed Baire measures on X.

Corollary 3 *Let $\{v_\alpha\}_{\alpha \in D}$ be a net in $M(X)$. Assume that for some $r \in R$ and all $\alpha \in D$, $\|v_\alpha\| \leq r$. Then the limit of $\{v_\alpha\}_{\alpha \in D}$ with respect to the weak* topology exists and is equal to v_1 if and only if for each infinite $\beta \in *D$ (i.e., $\beta \geq *\alpha$ for each standard $\alpha \in D$) we have $v_1 = (v_\beta)_X$.*

Proof The proof follows from Theorem 1 and the analog of Theorem II.A.4.9 for nets (Robinson [24, Theorem 4.2.5]).

Anderson and Rashid [3] have extended Corollary 3 to noncompact spaces and tight measures. A discussion of weak* cluster points and further applications of Theorem 1 can be found in [19]. An application that is a new standard result in potential theory, new even for the unit disk, is obtained in [18]. Henson [8] has shown that for any internal algebra $\mathcal{A}$ in $*X$ and $B \in \sigma(\mathcal{A})$, st($B$) is an analytic set. All analytic sets in X are obtained in this way if $\mathcal{A}$ is "rich" and each closed set in X is a G_δ.

We conclude this section with a construction of Lebesgue measure on $[0,1]$ using the standard part map; the first such construction was given by Anderson in [1]. It is well known that Lebesgue measure λ represents the Riemann integral as a functional on $C([0,1])$. Let δ_x denote unit mass at $x \in [0,1]$. For each $n \in N$, let

$$v_n = (1/n)\delta_0 + (1/n)\delta_{1/n} + \cdots + (1/n)\delta_{(1-n)/n}.$$

Then $\lim_{n \to \infty} \int f \, dv_n = \int_0^1 f(x) \, dx$ for each $f \in C([0,1])$, whence for $\omega \in *N - N$,

$$\lambda = (v_\omega)_{[0,1]}.$$

A similar construction using the measurable partitions defined in Section III yields Radon measures (see also Anderson [2]).

V. Applications to Stochastic Processes

A. Coin Tossing

Choose an $\omega \in {}^*N - N$ and let X be the set of all internal sequences of zeros and ones (or minus ones and ones) of length ω. We write $X = \{0,1\}^\omega$ (or $X = \{-1,1\}^\omega$). The internal cardinality of X, $|X|$, is $2^\omega \in {}^*N - N$. Let $\mathscr{A}$ be the class of internal subsets of X and for each $A \in \mathscr{A}$, let $v(A) = 2^{-\omega}|A|$. Then $(X, \mathscr{A}, v)$ is the internal probability space for the hyperfinite experiment of tossing a fair coin ω times. The space $(X, \sigma(\mathscr{A}), v_0)$ is a standard probability space that can be used as the basic space for ordinary infinite coin tossing as in [16].

Consider, for example, the internal event A_n for which the first $n-1$ tosses are tails and the nth toss is a head. If ω is even, then $A = \bigcup_{n=1}^{\omega/2} A_{2n} \in \mathscr{A}$ is the internal event of obtaining a head in ω tosses, the first occurring at an even numbered toss. Moreover,

$$v(A) = \sum_{n=1}^{\omega/2} \frac{1}{2^{2n}} = \frac{1}{3} - \frac{1}{3 \cdot 2^\omega}$$

and $v_0(A) = 1/3$. The standard event $B = \bigcup_{n=1, n \in N}^{\infty} A_{2n} \in \sigma(\mathscr{A})$ corresponds to getting the first head at an even numbered toss in an infinite number of tosses; $v_0(B) = 1/3$.

If $Y = \{0,1\}^N$ is the standard space of sequences of 0's and 1's, then by adjoining to each internal sequence in X an internally infinite sequence of 0's, we may assume that $X \subset {}^*Y$. If we restrict each internal sequence $x(i)$ in X to the standard natural numbers, we obtain the standard part of x with respect to the product topology. The measure v_Y defined in Section IV.B is the usual probability measure for coin tossing. In general, a proof of the Kolmogorov extension theorem can be obtained with a similar construction (see [19, Example 3]).

B. The Poisson Process

Let ω be an infinite factorial in ${}^*N$. That is, $\omega = 1 \cdot 2 \cdots (\eta - 1) \cdot \eta$ for some $\eta \in {}^*N - N$. For simplicity we choose a standard rational number λ as the parameter for our process, and we let γ be the infinite integer $\lambda \omega$. Divide the interval $[0, \omega]$ into ω^2 intervals $[0, 1/\omega), [1/\omega, 2/\omega), \ldots, [(\omega^2 - 1)/\omega, \omega)$, and let X be the internal set of all internal ways that γ "balls" can be put into the ω^2 "boxes" $[k/\omega, (k+1)/\omega)$. That is, X consists of internal sequences $\{x_i : 1 \leq i \leq \gamma\}$ with $1 \leq x_i \leq \omega^2$ for each i; clearly, $|X| = \omega^{2\gamma}$. Let $\mathscr{A}$ denote the set of all internal subsets of X; for each $A \in \mathscr{A}$, set $v(A) = |A|/\omega^{2\gamma}$. Following [16], we use the space $(X, \sigma(\mathscr{A}), v)$ as a standard probability space for the Poisson process; the rich internal structure of $(X, \mathscr{A}, v)$ yields easy calculations of standard probabilities.

Fix, for example, a $k \in N$, and let T be a finite interval of rational length t with rational endpoints. There are exactly $t\omega$ of the ω^2 intervals inside T, and the v-probability of any one of the γ balls being put in T is $t\omega/\omega^2 = t/\omega = \lambda t/\gamma$. Let A be the internal event "there are exactly k balls in T." Then

$$v(A) = \frac{\gamma!}{(\gamma-k)!k!} \cdot \left(\frac{\lambda t}{\gamma}\right)^k \cdot \left(1 - \frac{\lambda t}{\gamma}\right)^{\gamma-k}$$

$$= \frac{(\lambda t)^k}{k!} \cdot \frac{\gamma!}{\gamma^k(\gamma-k)!} \cdot \left(1 - \frac{\lambda t}{\gamma}\right)^\gamma \cdot \left(1 - \frac{\lambda t}{\gamma}\right)^{-k}$$

$$\simeq \frac{(\gamma t)^k}{k!} \left(1 - \frac{\lambda t}{\gamma}\right)^\gamma \simeq \frac{(\lambda t)^k}{k!} e^{-\lambda t} = v_0(A).$$

Of course, $v_0(A)$ is the value at k of the Poisson distribution with parameter λt.

Now given $x \in X$, we order the balls b_i by the order in which they fall in the line *R. Thus $b_i \leq b_{i+1}$ and $b_i = b_{i+1}$ if and only if b_i and b_{i+1} are in the same interval $[k/\omega, (k+1)/\omega)$. Again fix a positive standard rational number t, and fix $j > 0$ and $k \geq 0$ in N. Given $t_0 = 0, 1/\omega, 2/\omega, \ldots, (\omega^2 - 1)/\omega$, let C_{t_0} be the event "$b_j \in [t_0, t_0 + t)$," and let D_{t_0} be the event, "for $j + 1 \leq i \leq k$, $b_i \in [t_0, t_0 + t)$ and $b_{j+k+1} \notin [t_0, t_0 + t)$." Let $\gamma' = \gamma - j$. Given C_{t_0}, the conditional probability of getting a given ball of the remaining γ' balls in $[t_0, t_0 + t)$ is

$$\frac{t\omega}{\omega^2 - t_0\omega} = \frac{t}{\omega - t_0} = \frac{\lambda t}{\gamma - t_0\lambda} = \frac{\lambda t}{\gamma' + j - t_0\lambda}.$$

Therefore, for all finite t_0, and hence for all $t_0 \leq \tau$ for some infinite τ, the conditional probability

$$v(D_{t_0}|C_{t_0}) = \frac{\gamma'!}{(\gamma'-k)!k!} \cdot \left(\frac{\lambda t}{\gamma' + j - t_0\lambda}\right)^k \cdot \left(1 - \frac{\lambda t}{\gamma' + j - t_0\lambda}\right)^{\gamma'-k} \simeq \frac{(\lambda t)^k}{k!} e^{-\lambda t}.$$

On the other hand, $\sum_{t_0 < \tau} v(C_{t_0}) \simeq 1$, and so $\sum_{t_0 < \tau} v(D_{t_0}|C_{t_0}) \cdot v(C_{t_0}) \simeq (\lambda t)^k e^{-\lambda t}/k!$. That is, the v_0-probability of having exactly k more balls in the interval of length t after the jth ball is $(\lambda t)^k e^{-\lambda t}/k!$. This proves that for these internal stopping times, i.e., the time of the jth event, Poisson processes have the strong Markov property. Since

$$\sum_{k=0, k \in N}^{\infty} \frac{(\lambda t)^k}{k!} e^{-\lambda t} = e^{\lambda t} \cdot e^{-\lambda t} = 1,$$

the v_0-probability of having only a finite number of balls in any finite interval $[0, t)$ is 1. Moreover, since $\lim_{t \to 0} e^{-\lambda t} = 1$, the v_0-probability of having ball b_{j+1} infinitely close to b_j is 0, and since this is true for each $j \geq 1$ in N, it follows that the v_0-probability of having two balls in the same monad is 0.

Let $\mathscr{B}$ denote the Borel sets in the positive real numbers R^+. For each $t \in {}^*R^+$ and $x \in X$, let $f(t, x)$ be the number of balls in $[0, t]$, where we assume that any ball in the box $[k/\omega, (k+1)/\omega)$ is at k/ω. We would like to restrict the values of t to standard values and thus obtain standard sample paths. That we can do so is a consequence of the following general result from [16].

Theorem 1 *Let $\mathscr{A}$ be an internal σ-algebra (in the nonstandard sense) in the internal set X. Let $f : {}^*R^+ \times X \to {}^*R$ be an internal ${}^*\mathscr{B} \times \mathscr{A}$-measurable function such that $f(t, x)$ is an increasing function of t for each $x \in X$. Let $g : R^+ \times X \to R \cup \{+\infty - \infty\}$ be defined by setting $g(s, x) = \sup_{t \simeq s} {}^0 f(t, x)$ for each $x \in X$. Then $g(s, x)$ is an increasing and right continuous function of s for each $x \in X$, and g is $\mathscr{B} \times \sigma(\mathscr{A})$-measurable on $R^+ \times X$.*

Proof Clearly, $g(s, x)$ is an increasing function of s for each $x \in X$. Fix $x \in X$ and $s \in R^+$, and let $a = g(s, x)$. Assume that $g(\cdot, x)$ is not right continuous at s. Then there is an $\varepsilon > 0$ in R so that for each $n \in N$ there is a $t \in {}^*R$ with $s < t < s + (1/n)$ and $f(t, x) - a \geq \varepsilon$. By the permanence principle, there are an $\omega \in {}^*N - N$ and a t with $s < t < s + 1/\omega$ such that $f(t, x) - a \geq \varepsilon$. But ${}^0 f(t, x) \leq a$. Therefore, $g(\cdot, x)$ is right continuous at s.

To show that g is $\mathscr{B} \times \sigma(\mathscr{A})$-measurable, we let h_a be the function defined on X for a given $a \in R$ by setting $h_a(x) = \inf\{t \in {}^*R^+ : f(t, x) \geq a\}$. Then h_a is $\mathscr{A}$-measurable since $\{x : h_a(x) \geq 0\} = X$ and for any $\beta > 0$ in ${}^*R$

$$\{x : h_a(x) \geq \beta\} = \bigcap_{q \text{ rational in } {}^*R; \, q < \beta} \{x : f(q, x) < a\}.$$

It follows from Theorem IV.A.3 that ${}^0 h_a$ is $\sigma(\mathscr{A})$-measurable on X, and thus $\{(s, x) \in R^+ \times X : s < {}^0 h_a(x)\}$ is $\mathscr{B} \times \sigma(\mathscr{A})$-measurable. Thus for any $a \in R$, the following set is $\mathscr{B} \times \sigma(\mathscr{A})$-measurable:

$$\bigcup_{n=1, n \in N}^{\infty} \{(s, x) \in R^+ \times X : s < {}^0 h_{a-1/n}(x)\}$$

$$= \bigcup_{n=1, n \in N}^{\infty} \{(s, x) \in R^+ \times X : \forall t \simeq s, f(t, x) < a - 1/n\}$$

$$= \{(s, x) \in R^+ \times X : g(s, x) < a\}.$$

The mapping $f(\cdot, x) \to g(\cdot, x)$ defined by $g(s, x) = \sup_{t \simeq_s} f(t, x)$ is a standard part mapping. The relevant topology for the paths associated with the Poisson process is the Skorokhod J_1-topology, and for general internal increasing functions, it is the Skorokhod M_1-topology (see Skorokhod [26]).

C. Anderson's Construction of Brownian Motion and the Itô Integral

In [1], Anderson used the measure space construction of Section IV to obtain a representation of Brownian motion and a construction of the Itô integral. We give here a brief account of some of his results. First there is a nonstandard form of the central limit theorem. For this section, $0 \notin N$.

Theorem 1 (Anderson [1]) *Let $\{X_n : n \in {}^*N\}$ be an internal sequence of internally independent random variables on an internal measure space $(\Omega, \mathscr{A}, v)$. Assume that there is a standard distribution function F such that ${}^*F$ is the distribution of X_n, $E(X_n) = 0$, and $E(X_n^2) = 1$ for each $n \in {}^*N$. Let ψ denote the standard Gaussian distribution. Then for any $m \in {}^*N - N$ and any $\alpha \in {}^*R$,*

$$v\left(\left\{\omega \in \Omega : \frac{1}{\sqrt{m}} \sum_{n=1}^{m} X_n(\omega) \leqslant \alpha\right\}\right) \simeq {}^*\psi(\alpha).$$

The restriction that ${}^*F$ be a standard distribution function can be and was weakened in [1]. A Brownian motion can now be defined as follows.

Definition 2 Fix $\eta \in {}^*N - N$ and let $(\Omega = \{-1, 1\}^\eta, \mathscr{A}, v)$ be the internal space for coin tossing as in Section A. That is, $\mathscr{A}$ consists of all internal subsets of Ω and for each $A \in \mathscr{A}$, $v(A) = |A|/2^\eta$. Let $(\Omega, \mathscr{D}, P)$ be the v_0-completion of the standard probability space $(\Omega, \sigma(\mathscr{A}), v_0)$. For each $\omega \in \Omega$ and $k \leqslant \eta$, let $\omega_k = \omega(k) \in \{-1, 1\}$. Let χ denote the hyperfinite random walk on $(\Omega, \mathscr{A}, v)$ defined by setting

$$\chi(t, \omega) = \frac{1}{\sqrt{\eta}} \left[\sum_{i=1}^{[\eta t]} \omega_i + (\eta t - [\eta t])\omega_{[\eta t]+1}\right]$$

for each $t \in {}^*[0, 1]$ and $\omega \in \Omega$. Here $[\eta t]$ denotes the largest element of ${}^*N$ less than or equal to ηt. Let $\beta(t, \omega) = {}^0\chi(t, \omega)$ for each $(t, \omega) \in [0, 1] \times \Omega$.

Theorem 3 (Anderson [1]) *If $\eta \in {}^*N - N$, then β is a Brownian motion on $(\Omega, \mathscr{D}, P)$.*

Proof (i) Given $t \in [0, 1]$, $\chi(t, \cdot)$ is $\mathscr{A}$-measurable, whence by Theorem IV.A.3, $\beta(t, \cdot)$ is $\mathscr{D}$-measurable.

(ii) Given $s < t$ in $[0,1]$ and $\lambda = [\eta t] - [\eta s]$,

$$P(\{\omega \in \Omega : \beta(t,\omega) - \beta(s,\omega) \leq \alpha\})$$
$$= P(\{\omega : {}^0\chi(t,\omega) - {}^0\chi(s,\omega) \leq \alpha\})$$
$$= P\left(\left\{\omega : {}^0\sum_{k=[\eta s]}^{[\eta t]} \frac{\omega_k}{\sqrt{\eta}} \leq \alpha\right\}\right)$$
$$= \lim_{n \to \infty} v_0\left(\left\{\omega : \frac{1}{\sqrt{\lambda}} \sum_{k=[\eta s]}^{[\eta t]} \omega_k \leq \sqrt{\eta/\lambda}\left(\alpha + \frac{1}{n}\right)\right\}\right)$$
$$= \lim_{n \to \infty} {}^0({}^*\psi)\left(\sqrt{\frac{\eta}{\lambda}}\left(\alpha + \frac{1}{n}\right)\right)$$
$$= \lim_{n \to \infty} \psi\left({}^0\left(\sqrt{\frac{\eta}{\lambda}}\left(\alpha + \frac{1}{n}\right)\right)\right)$$
$$= \lim_{n \to \infty} \psi\left[\frac{\alpha + (1/n)}{\sqrt{t-s}}\right] = \psi\left[\frac{\alpha}{\sqrt{t-s}}\right].$$

Thus $P(\{\omega \in \Omega : \beta(t,\omega) - \beta(s,\omega) < \alpha\sqrt{t-s}\}) = \psi(\alpha)$, so $\beta(t,\omega) - \beta(s,\omega)$ has a normal distribution with mean 0 and variance $t - s$.

(iii) If $s_1 < t_1 \leq s_2 < t_2 \leq \cdots \leq s_n < t_n$ in $[0,1]$, then it is not hard to show that

$$\{\beta(t_1,\cdot) - \beta(s_1,\cdot),\ldots,\beta(t_n,\cdot) - \beta(s_n,\cdot)\}$$

are independent random variables on $(\Omega, \mathcal{D}, P)$ (see [1], p. 31]). It follows that β is a Brownian motion.

Let $C[0,1]$ denote the continuous real-valued functions on $[0,1]$. A function $g \in {}^*C[0,1]$ is called near-standard if for some $f \in C[0,1]$ we have $g \in m(f)$ with respect to the sup-norm topology; that is, $\sup_{t \in {}^*[0,1]}|{}^*f(t) - g(t)| \simeq 0$. If g is a near-standard function, then the function h defined by $h(s) = {}^0g(s)$ for each $s \in [0,1]$ is continuous on $[0,1]$ and $g \in m(h)$.

Theorem 4 (Anderson [1]) *There is a set $\Omega_0 \in \mathcal{D}$ with $P(\Omega_0) = 0$ such that for all $\omega \in \Omega - \Omega_0$, $\chi(\cdot,\omega)$ is near-standard. Thus $\beta(\ ,\omega)$ is a continuous function on $[0,1]$ for almost all $\omega \in \Omega$.*

Proof For each $m, n \in N$, let Ω_{mn} be the internal set given by

$$\Omega_{mn} = \left\{\omega \in \Omega : \text{for some } i < n, \sup_{t \in [i/n,(i+1)/n]} \chi(t,\omega) - \inf_{t \in [i/n,(i+1)/n]} \chi(t,\omega) > \frac{1}{m}\right\}.$$

Then for $\lambda = \eta/n + 1$,

$$v(\Omega_{mn}) \leqslant nv\left(\left\{\omega: \left(\sup_{t \in [0,1/n]} -\inf\right)\chi(t,\omega) > \frac{1}{m}\right\}\right)$$

$$\leqslant nv\left(\left\{\omega: \max_{1 \leqslant k \leqslant \lambda} \left|\sum_{1}^{k} \omega_i\right| > \frac{\sqrt{\eta}}{2m}\right\}\right)$$

$$\leqslant nv\left(\left\{\omega: \max_{1 \leqslant k \leqslant \lambda} \sum_{1}^{k} \omega_i > \frac{\sqrt{\eta}}{2m}\right\}\right) + nv\left(\left\{\omega: \min_{1 \leqslant k \leqslant \lambda} \sum_{1}^{k} \omega_i < -\frac{\sqrt{\eta}}{2m}\right\}\right)$$

$$\leqslant 2nv\left(\left\{\omega: \sum_{1}^{\lambda} \omega_i > \frac{\sqrt{\eta}}{2m}\right\}\right) + 2nv\left(\left\{\omega: \sum_{1}^{\lambda} \omega_i < -\frac{\sqrt{\eta}}{2m}\right\}\right)$$

$$= 4nv\left(\left\{\omega: \frac{1}{\sqrt{\lambda}}\sum_{1}^{\lambda} \omega_i > \frac{\sqrt{\eta/\lambda}}{2m}\right\}\right)$$

$$\simeq 4n^*\psi[\sqrt{\eta/\lambda}/(2m)] \simeq 4n\psi[\sqrt{n}/(2m)]$$

$$= (4n/\sqrt{2\pi}) \int_{\sqrt{n}/(2m)}^{\infty} e^{-t^2/2} \, dt.$$

For $\sqrt{n}/(2m) > 1$,

$$P(\Omega_{mn}) < 2n \int_{\sqrt{n}/2m}^{\infty} e^{-t/2} \, dt = 4ne^{-\sqrt{n}/(4m)}.$$

Let $\Omega' = \Omega - \bigcup_{m=1}^{\infty} \bigcap_{n=1}^{\infty} \Omega_{mn}$. Then

$$P(\Omega') = 1 - \sup_m \inf_n P(\Omega_{mn}) \geqslant 1 - \sup_m \inf_n 4ne^{-\sqrt{n}/(4m)} - 1$$

Fix $\omega \in \Omega$. If for some $t \in {}^*[0,1]$ we have ${}^0\chi(t,\omega) = +\infty$ or ${}^0\chi(t,\omega) = -\infty$, then $\omega \in \Omega_{mn}$ for all standard m and $n \in N$, whence $\omega \notin \Omega'$. If for some s and $t \in {}^*[0,1]$ with $s \simeq t$ we have ${}^0|\chi(s,\omega) - \chi(t,\omega)| = a > 0$, then for $m > 2/a$ we have $\omega \in \Omega_{mn}$ for all $n \in N$, whence $\omega \notin \Omega'$.

Now given $\omega \in \Omega'$, $\chi(t,\omega)$ is finite for all $t \in {}^*[0,1]$ and when $s \simeq t$ it follows that $\chi(s,\omega) \simeq \chi(t,\omega)$. By the permanence principle, Theorem II.B.2.8, $\beta(t,\omega)$ is continuous on $[0,1]$ and, moreover,

$$\sup_{t \in {}^*[0,1]} |{}^*\beta(t,\omega) - \chi(t,\omega)| \simeq 0.$$

In [1], Anderson shows that the mapping $\chi(\cdot,\omega) \to \beta(\cdot,\omega)$ from Ω' into $C[0,1]$ is measurable and thus induces a measure on $C[0,1]$ as in Section IV.B. The measure obtained is an extension of Wiener measure. The above construction also yields a simple proof in [1] of Donsker's theorem for Brownian motion.

Given the internal point set Ω, Anderson considers in [1] a function $f:[0,1] \times \Omega \to R$ that is Itô integrable in (the standard sense) with respect to the Brownian motion β defined on $[0,1] \times \Omega$. Fixing a hyperfinite partition of $*[0,1]$ by infinitesimal intervals A_i and letting $\mathscr{A}$ denote all internal unions of the A_i's, he then lifts f to an internal measurable function $g: *[0,1] \times \Omega \to *R$. That is, for almost all points $(t, \omega) \in *[0,1] \times \Omega$, $^0g(t,\omega) = f(^0t, \omega)$, where "almost all" means with respect to μ_0 with μ being the internal product of the nonstandard extension of Lebesgue measure and the internal measure v. For each $\omega \in \Omega$, $g(\cdot, \omega)$ is constant on each A_i since it is measurable with respect to $\mathscr{A}$. Therefore, the random walk χ is internally a function of bounded variation, and the internal Stieltjes integral of g with respect to χ exists.

Theorem 5 (Anderson [1]) *For each standard $t \in [0,1]$, the Itô integral $\int_0^t f(s, \omega) \, d\beta(s, \omega)$ is infinitely close to the value of the internal Stieltjes integral $\int_0^t g(\tau, \omega) \, d\chi(\tau, \omega)$. For v_0-almost all $\omega \in \Omega$, the internal function from $*[0,1]$ into $*C[0,1]$ defined by $t \to \int_0^t g(\tau, \omega) \, d\chi(\tau, \omega)$ is near-standard in $*C[0,1]$. Therefore, the map from $[0,1]$ into $C[0,1]$ defined by $t \to \int_0^t f(s, \omega) \, d\beta(s, \omega)$ is continuous for v_0-almost all $\omega \in \Omega$.*

Using the internal formula $(d\chi)^2 = dt$, Anderson has obtained an easy proof of Itô's lemma in [1], thus making exact the heuristic standard formula $(d\beta)^2 = dt$. Recently, using the methods discussed in this chapter along with other new results, H. J. Keisler has obtained a number of new standard results for stochastic differential equations.

REFERENCES

1. Anderson, R. M., A nonstandard representation for Brownian motion and Itô integration, *Israel J. Math.* **25** (1976), 15–46.
2. Anderson, R. M., Star-finite representations of measure spaces (to appear).
3. Anderson, R. M., and Rashid, S., A nonstandard characterization of weak convergence, *Proc. Amer. Math. Soc.* **69** (1978), 327–332.
4. Bernstein, A. R., and Loeb, P. A., A nonstandard integration theory for unbounded functions, *in* "Victoria Symposium on Nonstandard Analysis" (A. Hurd and P. Loeb, eds.), Lecture Notes in Mathematics, No. 369. Springer-Verlag, Berlin and New York, 1974, pp. 40–49.
5. Bernstein, A. R., and Wattenberg, F., Nonstandard measure theory, *in* "Applications of Model Theory to Algebra, Analysis, and Probability" (W. A. J. Luxemburg, ed.). Holt, New York, 1969, pp. 18–86.
6. Fichtenholz, G., and Kantorovitch, L., Sur les operations dans l'espace des fonctions bornées, *Studia Math.* **5** (1934), 69–98.
7. Henson, C. W., On the nonstandard representation of measures, *Trans. Amer. Math. Soc.* **172** (1972), 437–446.
8. Henson, C. W., Analytic sets, Baire sets, and the standard part map, *Canad. J. Math.* (to appear).
9. Henson, C. W., Unbounded Loeb measures, *Proc. Amer. Math. Soc.* (to appear).
10. Hersh, R., Brownian motion and nonstandard analysis, Tech. Rep. No. 277, Univ. of New Mexico, Albuquerque, 1973.

11. Hersh, R., and Greenwood, P., Stochastic differential and quasi-standard random variables, *in* "Conference on Probabilistic Methods in Differential Equations" (E. A. Pinsky, ed.), pp. 35–61. Lecture Notes in Mathematics, No. 451. Springer-Verlag, Berlin and New York, 1975.
12. Itô K., Stochastic integral, *Proc. Imp. Acad. Tokyo*, **20** (1944), 8.
13. Keisler, H. J., "Foundations of Infinitesimal Calculus." Prindle, Weber, & Schmidt, Boston, 1976.
14. Loeb, P. A., A nonstandard representation of measurable spaces, L_∞, and L_∞^*, *in* "Contributions to Nonstandard Analysis" (W. A. J. Luxemburg and A. Robinson, eds.). North-Holland Publ., Amsterdam, 1972, pp. 65–80.
15. Loeb, P. A., A nonstandard representation of Borel measures and σ-finite measures, *in* "Victoria Symposium on Nonstandard Analysis" (A. Hurd and P. Loeb, eds.), Lecture Notes in Mathematics, No. 369. Springer-Verlag, Berlin and New York, 1974, pp. 144–152.
16. Loeb, P. A., Conversion from nonstandard to standard measure spaces and applications in probability theory. *Trans. Amer. Math. Soc.* **211** (1975), 113–122.
17. Loeb, P. A., Applications of nonstandard analysis to ideal boundaries in potential theory, *Israel. J. Math.* **25** (1976), 154–187.
18. Loeb, P. A., A generalization of the Riesz–Herglotz Theorem on representing measures, *Proc. Amer. Math. Soc.* **71** (1978), 65–68.
19. Loeb, P. A., Weak limits of measures and the standard part map, *Proc. Amer. Math. Soc.* (to appear).
20. Luxemburg, W. A. J., A general theory of monads, *in* "Applications of Model Theory to Algebra, Analysis, and Probability" (W. A. J. Luxemburg, ed.). Holt, New York, 1969, pp. 18–86.
21. Parikh, R., and Parnes, M., Conditional probabilities and uniform sets, *in* "Victoria Symposium on Nonstandard Analysis" (A. Hurd and P. Loeb, eds.), Lecture Notes in Mathematics, No. 369. Springer-Verlag, Berlin and New York, 1974, pp. 180–194.
22. Rashid, S., Economies with Infinitely Many Traders, Ph.D. thesis, Yale Univ. New Haven, Connecticut, 1976.
23. Robinson, A., On generalized limits and linear functionals, *Pacific J. Math.* **14** (1964), 269–283.
24. Robinson, A., "Non-Standard Analysis." North-Holland Publ., Amsterdam, 1966.
25. Royden, H. L., "Real Analysis." Macmillan, New York, 1968.
26. Skorokhod, A. V., Limit theorems for stochastic processes, *Theory Probab. Appl.* **1** (1956), 261–290.
27. Stroyan, K. D., and Luxemburg, W. A. J., "Introduction to the Theory of Infinitesimals," Series on Pure and Applied Mathematics, No. 72. Academic Press, New York. 1976.
28. Yosida, K., and Hewitt, E., Finitely additive measures, *Trans. Amer. Math. Soc.* **72** (1952), 46–66.

AMS (MOS) 1980 Subject Classifications: 60A05, 60A10, 26E35, 60J65

Limit Theorems: Stochastic Matrices, Ergodic Markov Chains, and Measures on Semigroups*

ARUNAVA MUKHERJEA

DEPARTMENT OF MATHEMATICS
UNIVERSITY OF SOUTH FLORIDA,
TAMPA, FLORIDA

I.	Introduction and Preliminaries	143
II.	Limits of Convolutions in Groups and Semigroups: Analysis in Stochastic Matrices	147
	A. Random Walk on Two Lines	148
	B. Purity of Limits in Groups and Semigroups	154
	C. Purity Law in Stochastic Matrices	156
	D. A One-One Correspondence between Probability Measures with Two-Point Supports and Their Continuous Weak* Limits	159
III.	Ergodicity of Markov Chains and Probability Measures on Semigroups: An Interplay	163
	Appendix	181
IV.	Limit Theorems for Convolution Products of Probability Measures on Completely Simple Semigroups	183
	References	200

I. Introduction and Preliminaries

The theory of probability distributions on groups is now well developed and widely studied. Several books, including the book by Grenander [13], have appeared during the last fifteen years or so to cover different aspects of the theory. Numerous research papers in this area have been written

---

* This work is supported by an NSF Grant No. MCS 77-03639.

recently. The large body of references cited in [18] will verify our remark. Grenander's books [13, 14] thoroughly detail how in various mathematical and physical contexts one is confronted with convergence problems for convolution products of probability measures on groups and semigroups. Apart from being useful in various applications, the problems in this area have attracted and challenged many mathematicians to search for analogs of classical probability results on the line or circle. This paper, among other things, will study several such analogs. Here we will be mostly interested in semigroups, abstract and concrete. The problems in semigroups need to be attacked differently than those in groups. Many of these problems are hard while others give way to careful analysis. An account of some of these is given in [34]. This paper, while practically disjoint from [34] in content, will attempt to show the nature of some problems we have not previously considered and how certain semigroup results in this area can be applied. We do not strive to seek the greatest generality in presenting our results. We only propose to show what sort of results can be expected and suggest methods to deal effectively with these problems.

A substantial portion of our paper contains new results. Some known results (mostly with new or no proofs) have been included, however, to show their relevance to new results or as applications of other results in a different context. Thus, in Section III, we have considered problems on ergodicity of nonhomogeneous (and homogeneous) Markov chains to apply several of our results on measures on semigroups. We hope to have shown a useful interplay between these two apparently unrelated topics to demonstrate the feasibility of enriching each theory by an application of the other. In Section II, we consider limit theorems for probability measures on stochastic matrices by showing first how these problems come up naturally in studying random walks on simple geometric figures on the plane. In the last section we consider the abstract theory. Here we consider completely simple semigroups to study limit theorems for convolution products of probability measures, including generalizations of the well-known Paul Lévy theorem on the equivalence of convergence in probability, convergence with probability one, and convergence in distribution for sums of independent random variables.

Throughout this paper (unless otherwise stated), S will denote a locally compact Hausdorff second countable semigroup (i.e., an algebraic semigroup with locally compact Hausdorff topology and jointly continuous multiplication). (In some parts of the paper, local compactness plays no essential role and can be replaced by a metric assumption. The reader can verify this easily.)

By a measure on S, we will mean a finite regular nonnegative measure on the class of all Borel sets (generated by open sets) of S. A net of measures

μ_α is said to converge vaguely to a measure μ if for each $f \in C(S)$, the continuous functions with compact support, $\int f \, d\mu_\alpha \to \int f \, d\mu$. Sometimes, we will also use the term weak* convergence to mean vague convergence. When the measures μ_α and μ are all probability measures, weak* convergence will be referred to as weak convergence. In this case, for every bounded continuous function f, $\int f \, d\mu_\alpha \to \int f \, d\mu$. By the Banach–Alaoglu theorem, the set $B(S)$ of all measures μ with $\mu(S) \leqslant 1$ is compact in the weak* (or vague) topology. Also, $B(S)$ is an algebraic semigroup under the usual convolution operation of measures. We will write $\mu_1\mu_2$ to denote the convolution of μ_1 and μ_2. We will also write $\mu_{k,n}$ to denote $\mu_{k+1}\mu_{k+2}\cdots\mu_n$. For sets A and $B \subset S$ and any point x in S, we will use the following notations:

$$Ax^{-1} = \{y : yx \in A\}, \qquad x^{-1}A = \{y : xy \in A\};$$
$$AB^{-1} = \bigcup\{Ax^{-1} : x \in B\}, \qquad A^{-1}B = \bigcup\{x^{-1}B : x \in A\}.$$

For a measure μ, S_μ will always denote its support; and, $P(S)$ will denote the set of all probability measures on S. $P(S)$ is a topological semigroup in the weak* topology, while $B(S)$ is not even separately continuous in this topology with respect to convolution (as multiplication).

An important class of semigroups that have been widely studied by semigroupists and probabilists is the class of completely simple semigroups. For a detailed discussion of these semigroups, see [9, 34, 36]. For convenience, we include here certain basic facts about these semigroups needed for our discussions later on. A semigroup is called completely simple if it is simple and contains a primitive idempotent. When S is completely simple, it is well known that S is topologically isomorphic to the product structure $X \times G \times Y$, where

$$X \equiv E(Se), \qquad G \equiv eSe, \qquad Y \equiv E(eS).$$

Here $e = e^2$ is a fixed idempotent of S and $E(A)$ denotes the set of idempotents in A. The set X is a locally compact left zero semigroup, the set Y is a locally compact right zero semigroup, and G is a locally compact topological group. The multiplication in $X \times G \times Y$ is defined by

$$(x_1, g_1, y_1)(x_2, g_2, y_2) = (x_1, g_1 y_1 x_2 g_2, y_2).$$

It is known that when S is compact, then it has a kernel (i.e., a smallest two-sided ideal) K, which is completely simple. Also, the support of an idempotent probability measure on S is a completely simple semigroup. If S is a matrix semigroup (of $n \times n$ matrices over a field) and if S has a completely simple kernel, then the set of matrices of S of minimal rank is the kernel of S (see [8]). It follows that the kernel of a compact semigroup of real or complex matrices is the set of matrices of minimal rank.

In our investigations, the following theorem of Csiszár will be useful.

Theorem 1.1 *Let S be a compact semigroup and $x_1, x_2, \ldots$ an arbitrary sequence of elements of S. Write for $0 \leq k < l$,*

$$x_{k,l} = x_{k+1} x_{k+2} \cdots x_l.$$

Assume that $x \in S$ is an accumulation (or cluster) point of the sequence $x_{0,l} = x_1 x_2 \cdots x_l$. Then there is a subsequence (n_i) of positive integers such that for each positive integer k,

$$\lim_{i \to \infty} x_{k,n_i} = \bar{x}_k, \qquad \lim_{i \to \infty} \bar{x}_{n_i} = x_\infty$$

exist; here, $x_\infty = x_\infty^2$, and $\bar{x}_k = \bar{x}_k x_\infty$ for each k. These results remain true even when S is not compact, but S can be embedded in a compact Hausdorff space S' topologically; in this case, the above limits will exist, of course, in S'.

The next theorem that will be useful in our discussions is due to Rosenblatt [36].

Theorem 1.2 *Let S be a compact semigroup with kernel K. Suppose $\mu \in P(S)$ and that $S = \overline{\bigcup_{n=1}^\infty S_\mu^n}$. Then if G is any open set containing K,*

$$\lim_{n \to \infty} \mu^n(G) = 1.$$

The next two theorems are due to the author and will be relevant in our discussions (see [32] for their proofs).

Theorem 1.3 *Let $S (\equiv X \times G \times Y)$ be completely simple. Suppose $\mu \in P(S)$ and $S = \overline{\bigcup_{n=1}^\infty S_\mu^n}$. Then $\mu^n \to 0$ vaguely as $n \to \infty$ if and only if the group factor G is noncompact.*

Theorem 1.4 *Let S be a compact semigroup and $\mu \in P(S)$. Suppose $S = \overline{\bigcup_{n=1}^\infty S_\mu^n}$. Then the sequence μ^n converges weakly if and only if $\lim_{n \to \infty} \inf S_\mu^n$ is nonempty, where $\lim_{n \to \infty} \inf S_\mu^n = \{x \in S : \text{given any open set } V \text{ containing } x, \text{ there exists a positive integer } k \text{ such that for } n > k, V \cap S_\mu^n \text{ is nonempty}\}$.*

Note that one immediate consequence of Theorem 1.4 is the following: Suppose S is the compact semigroup of $n \times n$ stochastic matrices with usual topology and matrix multiplication. Let $\mu \in P(S)$ be such that the kernel K (consisting of all stochastic matrices with identical rows) intersects the closed semigroup generated by S_μ. Then μ^n converges weakly since $K \cap (\overline{\bigcup_{n=1}^\infty S_\mu^n}) \subset \lim_{n \to \infty} \inf S_\mu^n$. This inclusion is immediate since in this case, for $y \in S$ and $x \in K$, $yx = x$.

Our last two theorems in this section are again due to Csiszár [10]. Theorem 1.5 was also obtained by A. Tortrat in a slightly different form. This theorem is often called the Csiszár–Tortrat theorem.

Theorem 1.5 *Let S be a group and μ_n be a sequence in $P(S)$. Then either*

$$\sup_x \mu_{0,n}(Kx) \to 0 \quad \text{as} \quad n \to \infty$$

for every compact subset $K \subset S$ or there exist a sequence a_n of elements in S such that for all positive integers k, the sequence $\mu_{k,n} a_n$ converges weakly.

Theorem 1.6 *Let S be a group, and $X_1, X_2, \ldots$ be a sequence of independent random variables with values in S. Assume that the products $X_{k,n} = X_{k+1} \cdots X_n$ have limiting distributions as $n \to \infty$, for all $k \geq 0$. Then there exists a unique compact subgroup H such that all the above limiting distributions are H-uniform and the product $X_1 X_2 \cdots X_n$ converges mod H with probability one.*

In Section IV, our results will show what the analogs of the above two theorems should be for completely simple semigroups.

II. Limits of Convolutions in Groups and Semigroups: Analysis in Stochastic Matrices

In this section, we make certain assertions concerning purity and continuity of the weak* limit of convolution products of probability measures on groups. Similar questions can be asked in certain semigroups, where the limit behavior is quite different. To show this, we consider the semigroup of stochastic matrices. Among other things, we also consider the problem (the most interesting part of this section) of determining how two probability measures, whose convolution iterates have the same weak* limit, are related to each other. This is discussed in Subsection D.

In what follows, S is always a locally compact Hausdorff topological semigroup. In this section, Q's and greek letters λ, μ, etc., will denote probability measures on S, and X, Y, Z, W, T, etc., will stand for random variables on some probability space. We will write $Q_{k,n}$ to denote the convolution product $Q_{k+1} Q_{k+2} \cdots Q_n$. In this section, we will answer only very special cases of the following two general problems:

(a) Let Q_1 and Q_2 be in $P(S)$ such that the weak* limits of Q_1^n and Q_2^n are both the same probability measure. How are Q_1 and Q_2 related?

(b) Let (Q_n) be a sequence in $P(S)$ such that the weak* limit of $Q_{k,n}$ is $Q^1 \in P(S)$. When is Q^1 pure (i.e., purely discontinuous, continuous singular or absolutely continuous) with respect to some distinguished measure on S?

When S is a group, the weak convergence of Q_1^n means that the weak limit of Q_1^n is the normed Haar measure on the compact subgroup of S generated by the support S_{Q_1} of Q_1. Hence w*-lim Q_1^n = w*-lim Q_2^n here implies that S_{Q_1} and S_{Q_2} generate the same compact subgroup of S; here, of course, Q_1 can be quite different from Q_2. As we will show, the above asymptotic behavior is quite different in semigroups. In the semigroup of 2×2 stochastic matrices, we will show that w*-lim Q_1^n = w*-lim Q_2^n implies that $Q_1 = Q_2$, at least when Q_1 and Q_2 both have two-point supports. It is, of course, clear that if there are positive integers k and s such that $Q_1^k Q_2^s = Q_2^s Q_1^k$, then if w*-lim $Q_1^n = \lambda_1$, w*-lim $Q_2^n = \lambda_2$, where λ_1 and λ_2 are, respectively, *unique* solutions of $\lambda_1 Q_1 = \lambda_1$ and $\lambda_2 Q_2 = \lambda_2$, then it follows that $\lambda_1 \lambda_2 = \lambda_2 \lambda_1$, $(\lambda_2 \lambda_1) Q_1 = \lambda_2 \lambda_1$, and $(\lambda_1 \lambda_2) Q_2 = \lambda_1 \lambda_2$, so that $\lambda_1 = \lambda_2 \lambda_1 = \lambda_1 \lambda_2 = \lambda_2$. Also, we may observe that the above commutative property easily holds whenever both Q_1 and Q_2 can be expressed as convex combinations of iterates of some probability measure Q. Our final remark regarding problem (a) above is that if S is an abelian semigroup such that $S = \bigcup_{n=1}^{\infty} S_{Q_1^n} = \bigcup_{n=1}^{\infty} S_{Q_2^n}$, then if the sequence Q_1^n and Q_2^n both weakly converge, their limits must be the normed Haar measure on the unique compact subgroup of S, which is also the kernel (the smallest two-sided ideal) of S. So in this case also, Q_1 and Q_2 can be very much unrelated. [The author is thankful to A. Nakassis for many stimulating discussions concerning (a) and (b).]

Regarding problem (b), we will see that Q_k^1 = w*-lim$_{n \to \infty} Q_{k,n}$ is pure when S is a group. When S is a semigroup, this limit behavior is again different. We will show that if S is the semigroup of stochastic matrices (with usual topology), then w*-lim$_n Q^n$ is pure with respect to the Lebesgue measure on $[0,1]$ (identified with the kernel of S; but w*-lim $Q_{k,n}$ need not be pure for a sequence of nonidentical probability measures). For the sake of convenience, we divide this section into four subsections: A, B, C, and D. In Subsection A, we present some motivational discussion showing how the problems that we discuss in later subsections can arise in different contexts, such as number theory or random walks on stright lines. In Subsections B and C, we study purity problems for infinite convolutions in groups and semigroups and in particular, stochastic matrices. In Subsection D, we present an interesting result establishing a one-one correspondence between probability measures Q with two-point supports in 2×2 stochastic matrices and their continuous weak* limits.

A. Random Walk on Two Lines

To motivate our study in stochastic matrices, we consider a random walk on two lines on the plane:

$$y = (a_1 - b_1)x + b_1 \quad \text{and} \quad y = (a_2 - b_2)x + b_2.$$

LIMIT THEOREMS

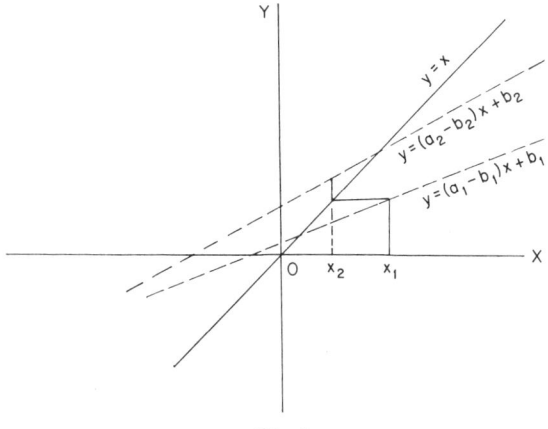

Fig. 1

A particle travels parallel to the Y-axis until it hits the line $y = (a_1 - b_1)x + b_1$ (with probability p) or the line $y = (a_2 - b_2)x + b_2$ (with probability q), where $p + q = 1$. After it hits one of the two lines, it travels parallel to the X-axis to hit the line $y = x$ (with probability 1) and then x_2 (see Fig. 1) becomes its new position at the beginning of the next step. Then the process is repeated. As shown in Fig. 1,

$$(x_1, 1 - x_1) \begin{pmatrix} a_1 & 1 - a_1 \\ b_1 & 1 - b_1 \end{pmatrix} = (x_2, 1 - x_2),$$

where $x_2 = (a_1 - b_1)x_1 + b_1$. It is clear from Fig. 1 that

$$(x_1, 1 - x_1) \begin{pmatrix} a_1 & 1 - a_1 \\ b_1 & 1 - b_1 \end{pmatrix} \begin{pmatrix} a_2 & 1 - a_2 \\ b_2 & 1 - b_2 \end{pmatrix} = (x_3, 1 - x_3),$$

etc. It follows that the value of x_n is given by the forward product of identically distributed random variables taking on the values

$$\begin{pmatrix} a_1 & 1 - a_1 \\ b_1 & 1 - b_1 \end{pmatrix} \quad \text{and} \quad \begin{pmatrix} a_2 & 1 - a_2 \\ b_2 & 1 - b_2 \end{pmatrix}$$

with probabilities p and q, respectively. The random variables $X_1, X_2, \ldots, X_n, \ldots$ are assumed independent and can be represented by

$$X_n = \begin{pmatrix} Y_n & 1 - Y_n \\ Z_n & 1 - Z_n \end{pmatrix} \quad (\equiv (Y_n, Z_n), \text{say}),$$

where (Y_n, Z_n) assumes the values (a_1, b_1) and (a_2, b_2) with probabilities p and q, respectively; also, (Y_n, Z_n) is independent of (Y_k, Z_k), $k \neq n$. Here the product $X_1 X_2 \cdots X_n$ has the same distribution as the backward product $X_n X_{n-1} \cdots X_1$ (the X_i's are identically distributed). Since the backward product gives us a more convenient (easy to handle) expression because of the way matrices multiply, we consider

$$X_n X_{n-1} \cdots X_1 \equiv (Y_{n,1}, Z_{n,1})$$

where

$$Y_{n,1} = Y_n(Y_{n-1} - Z_{n-1}) \cdots (Y_1 - Z_1)$$
$$+ \sum_{k=0}^{n-2} (Y_1 - Z_1)(Y_2 - Z_2) \cdots (Y_k - Z_k) Z_{k+1}$$

and

$$Z_{n,1} = Z_n(Y_{n-1} - Z_{n-1}) \cdots (Y_1 - Z_1)$$
$$+ \sum_{k=0}^{n-2} (Y_1 - Z_1)(Y_2 - Z_2) \cdots (Y_k - Z_k) Z_{k+1}$$

(In the above summation, the first term is Z_1.) Thus, we can write

$$y_n = (Y_{n,1} - Z_{n,1}) x_1 + Z_{n,1} = x_1 \prod_{i=1}^{n} (Y_i - Z_i) + Z_{n,1},$$

which has the same distribution as x_n. In what follows, we will study briefly the asymptotic behavior of y_n as $n \to \infty$. Let us write $W_i = Y_i - Z_i$: then we have

$$y_n = x_1 W_1 W_2 \cdots W_n + \sum_{k=1}^{n} W_1 W_2 \cdots W_{k-1} Z_k.$$

(The first term in this summation is Z_1.) When the lines are parallel, $a_1 - b_1 = a_2 - b_2 = d$, say; then

$$y_n = x_1 d^n + Z_1 + d Z_2 + \cdots + d^{n-1} Z_n.$$

So if $d < 1$, then

$$\lim_{n \to \infty} y_n = \sum_{k=0}^{\infty} d^k Z_{k+1},$$

which converges almost surely. For $d = 1$, we have

$$y_n = x_1 + Z_1 + Z_2 + \cdots + Z_n,$$

in which case the central limit theorem applies. For $d > 1$,
$$y_n/d^n = x_1 + (Z_1/d^n) + (Z_2/d^{n-1}) + \cdots + (Z_n/d),$$
which has the same distribution as
$$x_1 + (Z_1/d) + (Z_2/d^2) + \cdots + (Z_n/d^n)$$
and as $n \to \infty$ its distribution approaches the distribution of $x_1 + \sum_{i=1}^{\infty}(Z_i/d^i)$. It will follow from our subsequent discussions in this section that all these distributions are continuous.

Now suppose the lines are not parallel. Then $a_1 - b_1 \neq a_2 - b_2$. There exist real numbers s and t such that
$$b_1 = s(a_1 - b_1) + t \quad \text{and} \quad b_2 = s(a_2 - b_2) + t.$$
Then we have
$$s = \frac{b_1 - b_2}{(a_1 - b_1) - (a_2 - b_2)}, \quad t = \frac{a_1 b_2 - a_2 b_1}{(a_1 - b_1) - (a_2 - b_2)}.$$
It follows from the way the W_i's and Z_i's are distributed that
$$Z_i = sW_i + t.$$
Using this, we have
$$y_n = (s + t)[1 + W_1 + W_1 W_2 + \cdots + W_1 W_2 \cdots W_{n-1}] + (s + x_1) W_1 W_2 \cdots W_n - s.$$
If $s + t = 0$, then $b_1 - b_2 = -a_1 b_2 + a_2 b_1$ or $b_1/(1 - a_1) = b_2/(1 - a_2)$; this means that the points (a_1, b_1) and (a_2, b_2) are collinear on the plane with the point $(1, 0)$ (Fig. 2). As we will see later, this is the only case when the limiting distribution of $X_n X_{n-1} \cdots X_1$ is discrete and equal to the point mass at the matrix
$$\begin{pmatrix} b_1/(1 - a_1 + b_1) & (1 - a_1)/(1 - a_1 + b_1) \\ b_1/(1 - a_1 + b_1) & (1 - a_1)/(1 - a_1 + b_1) \end{pmatrix}.$$
Later on, in this section we show that the limiting distribution (even in the general case) is always continuous (in fact, either continuous singular or absolutely continuous), whenever it is not a point mass. In case $|a_1 - b_1|$ or $|a_2 - b_2|$ equals 1, it is easy to see that the limiting distribution of $X_n X_{n-1} \cdots X_1$ exists, unless the distribution of X_i is concentrated only at the matrix $\begin{pmatrix} 0 & 1 \\ 1 & 0 \end{pmatrix}$. In case $|a_1 - b_1| < 1$, $|a_2 - b_2| < 1$, the limiting distribution also exists and is given by the distribution of the almost surely convergent series
$$\lim_{n \to \infty} y_n = (s + t) \sum_{k=1}^{\infty} W_1 W_2 \cdots W_k + t.$$

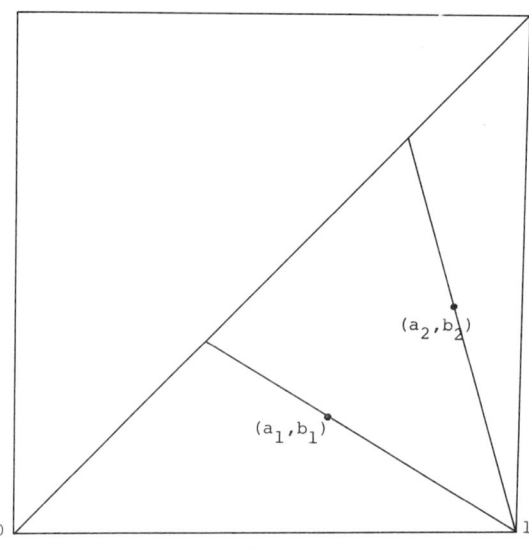

Fig. 2

The above problem was studied earlier in different contexts by Grenander [13], Rosenblatt [36], Maksimov [29], Sun [38], Grintsevichyus [15], and Mukherjea and Nakassis [33]. The problem of finding the nature of the limiting distribution as well as the distribution itself was studied by Mukherjea and Nakassis in the more general context of $n \times n$ stochastic matrices. For some detailed exposition on this and related aspects of the problem, the reader can consult [33, 34]. Actually, consideration of the problem in stochastic matrices is similar to its consideration in the context of the matrices

$$S_1 \equiv \left\{ \begin{pmatrix} a & 0 \\ b & 1 \end{pmatrix} : 0 \leqslant a, 0 \leqslant b, a+b \leqslant 1 \right\}.$$

Note that S_1 is isomorphic to the semigroup of stochastic matrices

$$S_2 \equiv \left\{ \begin{pmatrix} a & 1-a \\ b & 1-b \end{pmatrix} : 0 \leqslant b \leqslant a \leqslant 1 \right\}$$

under the isomorphism Φ defined by

$$\Phi : S_2 \to S_1, \quad \Phi(x) = AxA^{-1},$$

where $x \in S_2$,

$$A = \begin{pmatrix} 1 & -1 \\ 0 & 1 \end{pmatrix} \quad \text{and} \quad A^{-1} = \begin{pmatrix} 1 & 1 \\ 0 & 1 \end{pmatrix}.$$

Thus, we have

$$\Phi\begin{pmatrix} a & 1-a \\ b & 1-b \end{pmatrix} = \begin{pmatrix} a-b & 0 \\ b & 1 \end{pmatrix}, \quad \Phi\begin{pmatrix} a+b & 1-a-b \\ b & 1-b \end{pmatrix} = \begin{pmatrix} a & 0 \\ b & 1 \end{pmatrix}.$$

Maksimov and Grintsevichyus considered the problem in the group of matrices

$$\left\{ \begin{pmatrix} a & b \\ 0 & 1 \end{pmatrix} : a > 0 \right\}.$$

Grintsevichyus considered sums of the form

$$U_1 + U_2 V_1 + U_3 V_1 V_2 + \cdots + U_n V_1 V_2 \cdots V_{n-1} + \cdots,$$

where the U_i's and V_j's are real-valued random variables and investigated their almost sure convergence as n tends to infinity. He did not assume the independence of U_i and V_j for $i \neq j$. He proved the following theorem:

Theorem 2.1 *Let (V_n) be a sequence of independent identically distributed random variables, $V_1 \neq 0$ a.e. and $-\infty < E\log|V_1| < 0$. Suppose also that $U_1, U_2, \ldots$ is another sequence of independent identically distributed random variables. Then the series*

$$U_1 + U_2 V_1 + U_3 V_1 V_2 + \cdots$$

is almost surely convergent if and only if $E\log\sup(|U_i|, 1) < \infty$.

Maksimov, in his paper, is mostly concerned with finding the limiting distribution of Y_n as n tends to infinity. He also assumes that the distribution of X_n's has a two-point support. His paper [29] contains many interesting results in this case. Some new results are given in this context later in this section.

Before we close this discussion, we briefly digress to show a connection between the distribution of the series

$$T_1 + T_1 T_2 + \cdots + T_1 T_2 \cdots T_n + \cdots,$$

where the T_i's are uniformly distributed independent random variables on $(0, 1)$, and a problem in number theory. Let us define the function $f(x, y)$ as the number of integers less than or equal to x and free of prime factors greater than y. Chowla, Vidyaraghavan, Ramaswami, Buchstab, Dickman,[1] and deBruijn have shown that

$$\lim_{y \to \infty} \frac{f(y^t, y)}{y^t} = g(t),$$

---

[1] It seems that the result was first obtained by Dickman in *Ark. Mat. Astronom. Fys.* **22** (A10) (1930). It was rediscovered by others later.

where $g(t)$ is a function satisfying

$$tg'(t) + g(t-1) = 0, \quad t > 1,$$
$$g(t) = 0, \quad t < 0,$$
$$g(t) = 1, \quad 0 \leq t \leq 1.$$

Now let us write

$$T = \sum_{k=1}^{\infty} T_1 T_2 \cdots T_k, \quad T' = \sum_{k=2}^{\infty} T_2 T_3 \cdots T_k.$$

Then $T = T_1[1 + T']$; T_1, T' are independent. We will show that the density function of T is a constant multiple of $g(t)$ above.

To show this, let $F(t)$ be the distribution function of T. Then F is absolutely continuous and for $t > 0$,

$$F(t) = P(T \leq t) = P(T_1(1 + T') \leq t) = \int_0^1 F((t/x) - 1) \, dx.$$

Here note that T_1 and $1 + T'$ are independent and T, T' have the same distribution function. By a change of variable, we now have:

$$F(t) = t \int_{t-1}^{\infty} \frac{F(y)}{(1+y)^2} \, dy, \quad y = \frac{t}{x} - 1.$$

It is clear that $g_0(t) \equiv F'(t)$ exists for all $t > 0$ and

$$g_0(t) = (1/t)[F(t) - F(t-1)].$$

By differentiating again, we have for $t > 1$

$$tg_0'(t) = -g(t-1).$$

For $0 \leq t \leq 1$,

$$F(t) = t \int_0^{\infty} \frac{F(y)}{(1+y)^2} \, dy$$

so that for $0 \leq t \leq 1$,

$$g_0(t) = K, \quad K = \int_0^{\infty} \frac{F(y)}{(1+y)^2} \, dy.$$

This establishes the number theory connection with our random walk problem.

B. Purity of Limits in Groups and Semigroups

In this subsection, we consider questions of continuity and purity for the weak limit of the sequence $Q_{k,n}$ in groups and semigroups. The Brown

Moran paper [3] and the Hartman paper [16] are two good references on this topic. Let U be a class of Borel sets on S which is closed under countable unions and with the property that whenever A is an element of U, the translate Ax, x in S, is also an element of U. Then Q in $P(S)$ is called a pure probability measure if it has the following property with respect to every such class U: If $Q(A)$ is positive for some A in U, then $Q(B) = 1$ for some B in U. The measure Q is called continuous if $Q(x) = 0$ for every singleton x in S, and purely discontinuous if $Q(A) = 1$ for some countable set A. If m is in $P(S)$, then Q is called absolutely continuous with respect to m if $Q(B) = 0$ whenever $m(B) = 0$, and continuous singular with respect to m if it is continuous and $Q(B) = 1$ for some B with $m(B) = 0$. We may note that a pure probability measure on a group is always purely discontinuous or continuous singular or absolutely continuous (with respect to the Haar measure). (This follows easily by considering U as the class of at most countable sets or the class of m-null sets.) Following essentially the same kind of proof as in van Kampen's paper or Hartman's paper, assertions similar to those made in these papers regarding the purity nature of infinite convolutions can be made even in groups which are not necessarily abelian. We describe briefly a few of these results. The proofs run along the same lines as in the classical case.

Theorem 2.2 *Let $Q_n \in P(S)$ and w*-$\lim_{n \to \infty} Q_{0,n} = Q \in P(S)$. Then Q is discontinuous if $\prod_{n=1}^{\infty} \max_{x \in S} Q_n(x)$ is positive.*

Proof Choose x_n in S so that $\prod_{n=1}^{\infty} Q_n(x_n) = d$ (positive). Write $y_n = x_1 x_2 \cdots x_n$. Then $Q_{0,n}(y_n) \geq Q_1(x_1) Q_2(x_2) \cdots Q_n(x_n) \geq d$. This means that the sequence y_n must have a cluster point y and $Q(y) \geq d$. Q.E.D.

Lemma 2.1 *Suppose that S is a group and $Q_1 = Q_2 Q_3$ for Q_1, Q_2 and Q_3 in $P(S)$. Suppose also that for some x in S, $Q_1(x) = d > 0$; V_n is an open neighborhood of e such that $V_n = V_n^{-1}$ and $d \leq Q_1(xV_n^2) < d + (1/n)$ and $Q_3(V_n) > 1 - (1/n)$. Then there exist elements y_n, z_n in S such that $x = y_n z_n$, $z_n \in V_n$, and $|Q_2(y_n) - d| < 2/n$ and $Q_3(z_n) > 1 - [6/(nd)]$.*

The proof is omitted.
Now using Lemma 2.1, we can easily prove the following theorem.

Theorem 2.3 *Suppose that S is a group and $Q_n \in P(S)$. Suppose also that for every open neighborhood V of e, $\lim_k \inf_{n > k} Q_{k,n}(V) = 1$. Then $Q_{0,n}$ weakly converges to some Q in $P(S)$. If Q is discontinuous, then the infinite product $\prod_{n=1}^{\infty} \max_{x \in S} Q_n(x)$ is positive.*

We also present the following theorem on the purity of the limit (omitting the proof).

Theorem 2.4 *Suppose that S is a group and the Q_n's as in Theorem 2.3, satisfy the lim inf condition there. Suppose also that each Q_n is purely discontinuous. Then the limit measure Q is discrete or continuous singular or absolutely continuous.*

An easy example showing that Theorem 2.3 need not be valid even in compact semigroups is the following: Let $S = [0,1]$ under multiplication and usual topology. Choose Q_n to be any probability measure with its support contained in $[0, 1/n]$. Then it is clear that $S_{Q_{k,n}} \subset [0, 1/n]$ and $Q_{k,n}$ converges weakly to the unit mass at 0. So the limit here is discontinuous whereas the infinite product $\prod_{n=1}^{\infty} \max_{x \in S} Q_n(x)$ need not be positive. This argument extends easily to any compact semigroup with a zero. In the next subsection, we consider questions on the purity of the limit in the semigroup of stochastic matrices.

C. Purity Law in Stochastic Matrices

In this subsection, S will denote the set of all stochastic matrices of order two under multiplication and usual topology. Then S is a compact semigroup whose kernel K is given by

$$K = \left\{ \begin{pmatrix} a & 1-a \\ a & 1-a \end{pmatrix} : 0 \leq a \leq 1 \right\}.$$

We denote the matrix, whose first column elements are a and b, by the point (a, b). Let P be a probability measure on S. Suppose the sequence P^n converges weakly to a probability measure Q such that $S_Q \subset K$. Noting that K is a right zero semigroup, it follows easily that Q here is the unique solution of the convolution equation $\bar{X}P = \bar{X}$, where $\bar{X}$ is an element of $P(K)$.

Theorem 2.5 *Let P be a probability measure on S. Then the sequence P^n converges weakly to a probability measure if and only if P is not the unit mass at $(0, 1)$. If S_P contains some (a, b), where either $0 < a < 1$ or $0 < b < 1$, then P^n converges weakly to some $Q \in P(K)$.*

Proof Let S_1 be the compact semigroup generated by S_P, i.e., $S_1 = \overline{\bigcup_{n=1}^{\infty} S_P^n}$. Suppose that $S_1 \cap K \neq \emptyset$. Then $S_1 \cap K$ is the kernel of S_1. If P_1 is a cluster point of (P^n), then $S_{P_1} \subset S_1 \cap K$ [since by Theorem 1.2, $\lim_{n \to \infty} P^n(G) = 1 \; \forall$ open set $G \supset S_1 \cap K$.] This means that for any two cluster points P_1 and P_2, $P_2 = P_1 P_2 = P_2 P_1 = P_1$ (since $\forall xy \in K, xy = y$). This means that P^n is weakly convergent. Now suppose that $S_1 \cap K = \emptyset$. Then since

$$\det \begin{pmatrix} a & 1-a \\ b & 1-b \end{pmatrix} \neq 0 \quad \text{whenever} \quad a \neq b,$$

S_1 is a compact cancellative semigroup. By a well-known theorem of Numakura, S_1 is a group. By a theorem of Brown [2], S_1 is finite. Now notice that for $x = (a, b)$ in S_1, $x^n = (y_n, z_n)$, where

$$y_n = a(a - b)^n + b(a - b)^{n-1} + \cdots + b(a - b) + b,$$
$$z_n = b(a - b)^n + b(a - b)^{n-1} + \cdots + b(a - b) + b.$$

Since for $|a - b| < 1$, x^n converges to a point in K, $|a - b| = 1$. Hence, the points in S_1 (a finite group disjoint from K) are contained in

$$\left\{ \begin{pmatrix} 1 & 0 \\ 0 & 1 \end{pmatrix}, \begin{pmatrix} 0 & 1 \\ 1 & 0 \end{pmatrix} \right\}$$

(for $x = (a, b)$, $a \neq b$, $\lim x^n \in K \leftrightarrow x$ has a positive column).
The theorem now follows from the well-known Kawada–Ito theorem.
Q.E.D.

Theorem 2.6 *Suppose P is a probability measure on S, which is not the unit mass at $(0, 1)$. Then $Q = \text{w*-lim}_{n \to \infty} P^n$ exists. If S_Q has more than two points, then Q is continuous singular or absolutely continuous with respect to the Lebesgue measure on $[0, 1]$ (here we identify K with $[0, 1]$).*

Proof If P is not the unit mass at $(1, 0)$, then P^n converges weakly to some Q, whose $S_Q \subset K$. Then $QP = Q$. If all the points in S_P lie on a straight line containing $(1, 0)$, then it can be verified easily that this line contains the closed semigroup generated by S_P. Then the kernel of this semigroup is a single point and Q is a point mass. Suppose then that S_P contains two points, which are not collinear with $(1, 0)$. Then Q must be continuous. To prove this, let us assume that Q is discrete; then let $p = \sup\{Q(\{x\}) \mid x \in K\}$. Clearly this sup is attained and there exists u in K such that $Q(\{u\}) = p > 0$. Now we have

$$Q(u) = \int Q(ux^{-1}) P(dx) \quad \text{or} \quad \int [Q(u) - Q(ux^{-1})] P(dx) = 0.$$

Noting that ux^{-1} is a singleton

$$\left(\frac{u - x_2}{x_1 - x_2}, \frac{u - x_2}{x_1 - x_2} \right) \quad \text{for} \quad x = (x_1, x_2),$$

it is clear that $Q(ux^{-1}) = Q(u) = p$ for almost all $x(P)$. By the upper semicontinuity of the mapping $x \to Q(ux^{-1})$, it follows that $Q(ux^{-1}) = Q(u)$ for all $x \in S_P$. Let $y = (y_1, y_2)$ and $z = (z_1, z_2)$ be two points in S_P which are not collinear with $(1, 0)$. Then it follows that either uy^{-1} or uz^{-1} is different from u since $u = ux^{-1}$ if and only if u, x and $(1, 0)$ are collinear. Suppose $uy^{-1} \neq u$. Then the sequence (uy^{-n}) is infinite, which is a contradiction since $Q(S) = 1$. Hence, Q is a continuous measure.

Now we show that Q is either continuous singular or absolutely continuous. Using the Lebesgue decomposition theorem, we write

$$Q = \alpha Q_1 + \beta Q_2 \qquad (Q = QP),$$

where $0 \leqslant \alpha, 0 \leqslant \beta, \alpha + \beta = 1, Q_1 \perp m$, and $Q_2 \ll m$. (Here m is the Lebesgue measure on $[0,1]$.) Now note that if $I = [a,b] \subset [0,1]$, then for $x = (x_1, x_2)$, $Ix^{-1} = [ax^{-1}, bx^{-1}] \cap [0,1]$ and $m(Ix^{-1}) \leqslant 1/(x_1 - x_2) \cdot m(I)$. This implies that for any Borel set $A \subset [0,1]$, $m(A) = 0$ implies that $m(Ax^{-1}) = 0$. Hence $Q_2 P \ll m$, since

$$Q_2 P(A) = \int Q_2(Ax^{-1})\mu(dx).$$

Consider now $QP = \alpha Q_1 P + \beta Q_2 P$. Suppose $\beta > 0$, $\alpha > 0$. Then writing $Q_1 P = \gamma_1 Q_{11} + \gamma_2 Q_{12}$, where $\gamma_1, \gamma_2 \geqslant 0, \gamma_1 + \gamma_2 = 1, Q_{11} \perp m$, and $Q_{12} \ll m$. Hence it follows that $\alpha Q_1 - \alpha_1 \gamma_1 Q_{11} = \alpha \gamma_2 Q_{12} + \beta Q_2 P - \beta Q_2 = 0$. Hence $\gamma_1 = 1$, so that $Q_1 P = Q_{11} \perp m$. Now $\alpha Q_1 + \beta Q_2 = \alpha Q_1 P + \beta Q_2 P$, implying that $Q_1 \mu = Q_1, Q_2 \mu = Q_2$. By the uniqueness of the solution of $\bar{X}P = \bar{X} \in P(K)$, it follows that $Q_1 = Q_2 = Q$, which is a contradiction. Hence either $\alpha = 0$ or $\beta = 0$. Q.E.D.

The purity character of the limit in Theorem 2.6 may not hold if we consider a weakly convergent sequence $Q_{0,n} = Q_1 Q_2 \cdots Q_n$, where the Q_i's are not necessarily identical probability measures. The following example demonstrates this.

Example 2.1 Consider the kernel K (which is a right zero semigroup) of the semigroup S of stochastic matrices. We identify K with $[0,1]$. Define the measures Q_n by

$$Q_n(\{0\}) = 1/2 \qquad \forall n \geqslant 1,$$
$$Q_1(\{1/2\}) = Q_1(\{3/4\}) = 1/4,$$
$$Q_2(\{1/2\}) = Q_2(\{5/8\}) = Q_2(\{3/4\}) = Q_2(\{7/8\}) = 1/8,$$
$$Q_3(\{1/2\}) = Q_3(\{(1/2) + (1/16)\}) = Q_3(\{5/8\}) = Q_3(\{(5/8) + (1/16)\})$$
$$= Q_3(\{3/4\}) = Q_3(\{(3/4) + (1/16)\}) = Q_3(\{7/8\})$$
$$= Q_3(\{7/8\} + 1/16) = 1/16,$$

and so on.

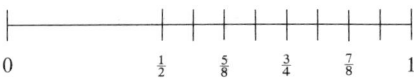

Then since $K (= [0,1])$ here is a right zero semigroup, $Q_{0,n} = Q_1 Q_2 \cdots Q_n = Q_n$.

Notice that for all x in $[0,1]$, $Q_n([0,x))$ is an increasing sequence of functions. Then we define Q' on open intervals of $[0,1]$ by

$$Q'((a,b)) = \lim_{n\to\infty} Q_n((a,b)).$$

The above limit exists since $Q_n((a,b)) = Q_n([0,b)) - Q_n([0,a))$. Now we can uniquely extend Q' to a probability measure on the Borel subsets of $[0,1]$ using the Caratheodory extension theorem of measure theory. We claim the following:

(i) $Q_n \to Q'$ as $n \to \infty$ in the weak* sense;
(ii) $Q'(\{0\}) = 1/2$, but Q' is continuous on $[1/2, 1]$.

To prove (i), let Q'' be any cluster point of the sequence (Q_n). Then $Q''(G) \leq \underline{\lim} Q_{n_i}(G)$, for some subsequence (Q_{n_i}) and open set G, which implies that $Q''(G) \leq Q'(G)$ for every open G. This means that for any closed set F, $Q''(F) \geq Q'(F)$. By regularity of the measures Q' and Q'', it follows that $Q' = Q''$. Thus (i) is proved. To prove (ii), let $1/2 \leq x \leq 1$. Now let $\varepsilon > 0$ and N be a positive integer such that $1/2^N < \varepsilon$. Choose an open interval I such that $Q_N(I) < \varepsilon$ and $x \in I$. Clearly, for $n > N$, $Q_n(I) < \varepsilon$ so that $Q'(I) \leq \underline{\lim}_{n\to\infty} Q_n(I) < \varepsilon$, implying that $Q'(\{x\}) = 0$ for $1/2 \leq x \leq 1$. This proves (ii).

D. A One-One Correspondence Between Probability Measures with Two-Point Supports and Their Continuous Weak* Limits

In this subsection, we will consider two probability measures P and Q, each with two-point support in the semigroup S of stochastic matrices such that both the sequences P^n and Q^n weakly converge to the same probability measure P^Q with support contained in the kernel K of S. We will show that P is necessarily equal to Q. The work in this subsection is joint work with A. Nakassis.

Suppose $A(x_1, y_1)$ and $B(x_2, y_2)$, where $x_1 > y_1$ and $x_2 > y_2$, are the points of S_P and

$$P(\{(x_1, y_1)\}) = p, \qquad P(\{(x_2, y_2)\}) = q.$$

Let $A'(x'_1, y'_1)$ and $B'(x'_2, y'_2)$, where $x'_1 > y'_1$ and $x'_2 > y'_2$, be the points of S_Q and

$$Q(\{(x'_1, y'_1)\}) = p' \quad \text{and} \quad Q(\{(x'_2, y'_2)\}) = q'.$$

Let P^n and Q^n both w*-converge to P^Q. Suppose G is the distribution function corresponding to P^Q on K (identified with $[0,1]$). Note that

$$G(x) = pG\left(\frac{x-y_1}{x_1-y_1}\right) + qG\left(\frac{x-y_2}{x_2-y_2}\right)\cdots. \qquad (2.1)$$

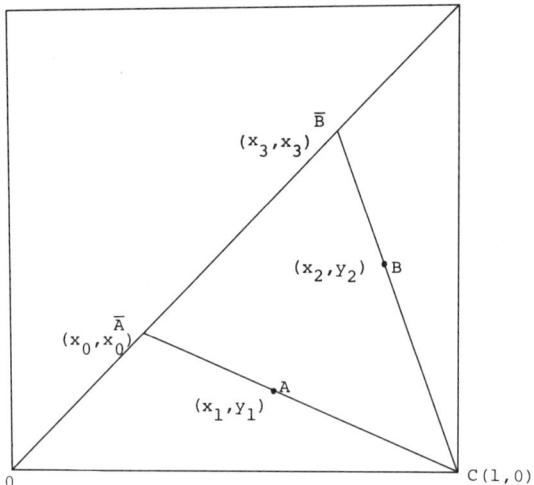

Fig. 3

[Here, we have

$$(x, x)(x_1, y_1)^{-1} = \left(\frac{x - y_1}{x_1 - y_1}, \frac{x - y_1}{x_1 - y_1}\right).\right]$$

In Fig. 3,

$$\frac{y_1}{1 - (x_1 - y_1)} = x_0 = \frac{x_0 - y_1}{x_1 - y_1}, \quad x_3 = \frac{x_3 - y_2}{x_2 - y_2}\left(= \frac{y_2}{1 - (x_2 - y_2)}\right);$$

also all the stochastic matrices represented by points of the triangle $\overline{A}\overline{B}C$ is a compact semigroup. It is then clear that $G(x_0) = 0$, $G(x_3) = 1$. Now we wish to write equation (2.1) in a more convenient form. So we write

$$L = x_0 - \frac{x_0 - y_2}{x_2 - y_2}, \quad a = \frac{1}{x_1 - y_1}, \quad \text{and} \quad b = \frac{1}{x_2 - y_2}.$$

Then defining g by $g(x) = G(Lx + x_0)$, we have equation (2.1) in the following form:

$$g(x) = pg(ax) + qg(bx - 1) \cdots . \tag{2.2}$$

Taking $c = 1/(b - 1)$, the equation (2.2) becomes

$$g(x) = pg(ax) + qg(b(x - c) + c) \cdots . \tag{2.3}$$

Now we claim the following:

(i) $g(c) = 1$ and for $x < c$, $g(x) < 1$;
(ii) $g(0) = 0$ and for $x > 0$, $g(x) > 0$. $\qquad(2.4)$

To prove (i), notice that $Lc + x_0 = x_3$ and so

$$g(c) = G(Lc + x_0) = G(x_3) = 1.$$

Suppose now that there exists $x_0 < c$ such that $g(x_0) = 1$. Then let

$$u = \inf\{x \mid x < c, g(x) = 1\}$$

so that $0 < u < c$. Since g is a continuous function, $1 = g(u)$; also,

$$g(u) = pg(au) + qg(bu - 1).$$

Since $au > u$, $g(bu - 1) = 1$. But since $u < c = 1/(b-1)$, $bu - 1 < u$. This contradicts the minimality of u. A similar argument establishes (ii) above. Now let $S_Q = \{(x_1', y_1'), (x_2', y_2')\}$ and w*-$\lim_{n \to \infty} Q^n = P^Q$. Let $Q(x_1', y_1') = p'$ and $Q(x_2', y_2') = q'$, $p' + q' = 1$. It is clear from the picture and the properties (2.4) that one of the points of S_Q must lie on the line CA and the other one on the line CB. So we assume with no loss of generality that (x_1', y_1') lies on CA and (x_2', y_2') lies on CB. This means that

$$\frac{y_1}{1 - x_1} = \frac{y_1'}{1 - x_1'}, \quad \frac{y_2}{1 - x_2} = \frac{y_2'}{1 - x_2'} \cdots . \tag{2.5}$$

We now show that the function g also satisfies the following equation:

$$g(x) = p'g(a'x) + q'g(b'(x - c) + c) \cdots , \tag{2.6}$$

where $a' = 1/(x_1' - y_1')$ and $b' = 1/(x_2' - y_2')$.

Notice that the function G also satisfies the equation

$$G(x) = p'G\left(\frac{x - y_1'}{x_1' - y_1'}\right) + q'G\left(\frac{x - y_2'}{x_2' - y_2'}\right) \cdots . \tag{2.7}$$

Writing $h(x) = G(L'x + x_0')$, where $L' = x_0' - [(x_0' - y_2')/(x_2' - y_2')]$ and $x_0' = y_1'/[1 - (x_1' - y_1')]$ $[= x_0$, by (2.5)$]$, $b' = 1/(x_2' - y_2')$, $a' = 1/(x_1' - y_1')$, and $c' = 1/(b' - 1)$, we have (after some computations)

(1) $L/(b - 1) = L'/(b' - 1)$ (though L may not equal L'),
(2) $h(c'x) = g(cx) \quad \forall x \in [0, 1]$.

Also using (2.7), we have

$$h(x) = p'h(a'x) + q'h(b'(x - c') + c').$$

This means that

$$h(c'x) = p'h(a'c'x) + q'h(b'(c'x - c') + c')$$

or

$$g(cx) = p'g(a'cx) + q'g(c[b'(x - 1) + 1])$$

Using y for cx, we have
$$g(y) = p'g(a'y) + q'g(b'(y-c) + c),$$
proving (2.6). Note that it is not clear at all if $c = c'$.

We will now study equations (2.3) and (2.6). From (2.3) and (2.6), we have
$$g(x) = pp'g(a'ax) + pq'g(b'(ax - c) + c) \\ + qp'g(a'b(x-c) + a'c) + qq'g(b'b(x-c) + c) \tag{2.8}$$
and
$$g(x) = pp'g(aa'x) + p'qg(b(a'x - c) + c) \\ + q'pg(ab'(x-c) + ac) + q'qg(bb'(x-c) + c). \tag{2.9}$$

Hence from (2.8) and (2.9), we have
$$pq'g(b'(ax - c) + c) + qp'g(a'b(x-c) + a'c) \\ = p'qg(b(a'x - c) + c) + q'pg(ab'(x-c) + ac) \tag{2.10}$$

Notice that the left-hand side is $pq' + qp'$ for the first time when $x = \max\{c/a, c - (c/b) + [c/(a'b)]\}$. Now we claim that

$$\frac{1}{a} \leq 1 - \frac{1}{b} + \frac{1}{a'b} \quad \text{and} \quad \frac{1}{a'} \leq 1 - \frac{1}{b'} + \frac{1}{ab'} \tag{2.11}$$

We only prove one of the inequalities. The proof of the other is similar. Suppose that $1/a' > 1 - (1/b') + 1/ab'$. Put $x = (c/a')(<c)$ in (2.10). Then the right-hand side of (2.10) $> p'q + pq'$ since $ab'[(c/a') - c] + ac > c$. Therefore, because of (2.4), we must have
$$a'b[(c/a') - c] + a'c \geq c,$$
which is impossible since
$$(1/a') - 1 + (1/b) - [1/(a'b)] = [(1/a') - 1][1 - (1/b)] < 0$$
(remember: $a' > 1, b > 1$).

Because of (2.11), it now follows that

$$c - \frac{c}{b} + \frac{c}{a'b} = c - \frac{c}{b'} + \frac{c}{ab'} \quad \text{or} \quad \frac{1}{b}\left(1 - \frac{1}{a'}\right) = \frac{1}{b'}\left(1 - \frac{1}{a}\right) \tag{2.12}$$

Similarly, checking again equation (2.10) we have
$$\min\{(c/a) - [c/(ab')], c - (c/b)\} = \min\{(c/a') - [c/(a'b)], c - (c/b')\}. \tag{2.13}$$

Notice that
$$(c/a) - [c/(ab')] - [c - (c/b')] = c[(1/a) - 1][1 - (1/b')] \neq 0;$$

also
$$(c/a') - [c/(a'b)] - [c - (c/b)] = c[(1/a') - 1][1 - (1/b)] \neq 0.$$
Hence it follows from (2.13) that *either*
$$c - (c/b) = c - (c/b'),$$
in which case $b = b'$ and then by (2.12), $a = a'$; or
$$(c/a) - [c/(ab')] = (c/a') - [c/(a'b)],$$
in which case we have
$$(1/a)[1 - (1/b')] = (1/a')[1 - (1/b)]. \tag{2.14}$$
From (2.12) and (2.14),
$$(1/a) - (1/b') = (1/a') - (1/b). \tag{2.15}$$
But again by (2.12) and (2.15),
$$\frac{(1/b)}{(1/b')} = \frac{1 - (1/a)}{1 - (1/a')} = \frac{1 - (1/a) + (1/b)}{1 - (1/a') + (1/b')} = 1$$
so that $b = b', a = a'$.
Hence we have:
$$x_1 - y_1 = x'_1 - y'_1, \qquad x_2 - y_2 = x'_2 - y'_2.$$
Since
$$\frac{y_1}{1 - x_1} = \frac{y'_1}{1 - x'_1}, \qquad \frac{y_2}{1 - x_2} = \frac{y'_2}{1 - x'_2},$$
we have:
$$\frac{y_1}{1 - (x_1 - y_1)} = \frac{y'_1}{1 - (x'_1 - y'_1)} \quad \text{and} \quad \frac{y_2}{1 - (x_2 - y_2)} = \frac{y'_2}{1 - (x'_2 - y'_2)}.$$
It follows that $y_1 = y'_1, y_2 = y'_2, x_1 = x'_1, x_2 = x'_2$. It is now clear that $p = p'$, $q = q'$, and therefore, $P = Q$.

III. Ergodicity of Markov Chains and Probability Measures on Semigroups: An Interplay[2]

Convergence problems for products of stochastic matrices naturally come up in various contexts in biology, economics, and various other applications.

---

[2] The author is thankful to Dean Isaacson and Eugene Seneta for some comments on a part of this section. (See Note 1 added in proof at end of paper.)

To take one specific example, we discuss here the problem of tendency to consensus in an information exchanging operation. This problem can be set up suitably (but not artificially) to show its connection with the problem of ergodicity for products of stochastic matrices.

The consensus problem can be put simply as follows: A number of individuals acting as a group have to estimate an unknown parameter (or probability). Each individual in the group has his own estimate or an idea of a probability distribution for the parameter. But when he knows of the other estimates by other individuals in the group, he modifies his own estimate. So the problem is whether a single distribution can be arrived at with the agreement of every member in the group. To be more precise, suppose there are n individuals in the group with their initial estimates $E_0 = (E_1^0, E_2^0, \ldots, E_m^0)$. Suppose that the ith individual wishes to revise his own estimate to take into account the estimates of others by attaching an initial weight $P_{ij}^{(1)}$ to the estimate of the jth individual. Then his revised estimate after the first step information exchange becomes

$$E_i^1 = \sum_{j=1}^m P_{ij}^{(1)} E_j^0,$$

where $\sum_{j=1}^n P_{ij}^{(1)} = 1$ for all i, $1 \leq i \leq m$. Let us denote the $m \times m$ stochastic matrix $(p_{ij}^{(1)})$ by P_1. Let P_k denote the stochastic matrix $(p_{ij}^{(k)})$, where $p_{ij}^{(k)}$ is the weight that the ith individual attaches to the $(k-1)$th step revised estimate of the jth individual to revise his own estimate at the kth information exchange. Thus, we have

$$E_k = (E_1^k, E_2^k, \ldots, E_m^k),$$
$$E_k = P_k P_{k-1} \cdots P_2 P_1 E_0. \quad (3.1)$$

In the simplest case, when all the P_i's are the same and each equals P, then

$$E_k = P^k E_0.$$

Clearly then the consensus problem is the problem of the existence of $\lim_{n \to \infty} P^k$. The homogeneous case above was considered by DeGroot [11] and the nonhomogeneous case by Chatterjee and Seneta [7]. One interesting feature of equation (3.1) is that here the matrix product is a backward product rather than a forward product as one comes across in nonhomogeneous Markov chain theory. Here we will point out an essential difference in behavior between the backward and forward products and then discuss the above convergence problem for matrix products for the special class of bistochastic matrices, using the convergence theory for convolution products of probability measures on finite groups.

First we need some definitions. We write $P_{k,n} = (p_{ij}^{k,n})$ to denote $P_{k+1} P_{k+2} \cdots P_n$ and $P'_{k,n} = (p_{ij}'^{k,n})$ to denote $P_n P_{n-1} \cdots P_{k+1}$.

Definition 3.1 The sequence of matrices (P_i) is weakly ergodic for forward products (respectively, for backward products) if for each i, j, l and k,

$$\lim_{n \to \infty} [p_{ij}^{k,n} - p_{lj}^{k,n}] = 0, \quad \text{respectively,} \quad \lim_{n \to \infty} [p_{ij}'^{k,n} - p_{lj}'^{k,n}] = 0).$$

Definition 3.2 A weakly ergodic sequence of matrices (P_i) is called strongly ergodic for forward products (respectively, for backward products) if for all i, j, and k,

$$\lim_{n \to \infty} p_{ij}^{k,n} \quad (\text{respectively,} \lim_{n \to \infty} p_{ij}'^{k,n})$$

exists.

It is, of course, obvious that strong ergodicity implies weak ergodicity. But the converse implication for forward products is not valid. For example, if

$$P_n = \begin{cases} \begin{bmatrix} 1 - (1/n) & 1/n \\ 1 - (1/n) & 1/n \end{bmatrix} & \text{if } n \text{ is even} \\ \begin{bmatrix} 1/n & 1 - (1/n) \\ 1/n & 1 - (1/n) \end{bmatrix} & \text{if } n \text{ is odd,} \end{cases}$$

then $P_k P_{k+1} \cdots P_n = P_n$ and $\lim_{n \to \infty} P_n$ doesn't exist, though the chain is obviously weakly ergodic for forward products. But for backward products, $P_n P_{n-1} \cdots P_k = P_k$ and therefore, $\lim_{n \to \infty} P'_{k,n} = P_k$ so that the chain is strongly ergodic in this case. Indeed, the following holds:

Theorem 3.1 *For backward products, weak and strong ergodicity are equivalent.*

Proof Since $P'_{k,n+p} = P'_{n,n+p} P'_{k,n}$, we have

$$p_{ij}'^{k,n+p} = \sum_l p_{il}'^{n,n+p} p_{lj}'^{k,n}$$

By weak ergodicity, given $\varepsilon > 0$, there exists a positive integer N such that $n > N$ implies

$$\varepsilon < -p_{ij}'^{k,n} + p_{lj}'^{k,n} < \varepsilon$$

for all i, l, and j. This means that

$$[p_{ij}'^{k,n} - \varepsilon] \sum_l p_{il}'^{n,n+p} < p_{ij}'^{k,n+p} < [p_{ij}'^{k,n} + \varepsilon] \sum_l p_{il}'^{n,n+p}$$

or

$$|p_{ij}'^{k,n} - p_{ij}'^{k,n+p}| < \varepsilon \quad (\text{for all } p).$$

From the Cauchy property, strong ergodicity is immediate. Q.E.D.

Theorem 3.1 (first given in [7] with a slightly different proof) is quite elementary. However, the key property[3] of matrices that worked there is that $Q_1 Q_2 = Q_2$ whenever Q_2 is a stochastic matrix with identical rows. In other words, each such matrix is a right zero in the compact semigroup of stochastic matrices (with usual topology and matrix multiplication). Thus we can also give a very simple compactness proof of the above theorem, which also will show why the theorem should not be valid for forward products. Let Q_1 and Q_2 be limit points of the sequence $P'_{k,n}$. Then by the weak ergodicity assumption, Q_1 and Q_2 are matrices with identical rows; also there exist sequences n_i, n'_i ($n_i + < n'_i$) of positive integers such that $P'_{k,n_i} \to Q_1$, and $P'_{k,n'_i} \to Q_2$ as $i \to \infty$. Then $P'_{k,n'_i} = P'_{n_i,n'_i} P'_{k,n_i}$. If S' is a limit point of the sequence (P'_{n_i,n'_i}), by the joint continuity of multiplication it follows that $Q_2 = S' \cdot Q_1$. But $S' \cdot Q_1 = Q_1$. Thus the theorem is proved again.

For forward products of bistochastic matrices, it is immediate that weak and strong ergodicity are equivalent. For forward products of stochastic matrices, weak ergodicity coupled with the existence of limits of all diagonal entries of $P^{k,n}$ (for all k) implies strong ergodicity. This follows immediately from the definition. A number of interesting results giving necessary and sufficient conditions for the ergodicity of backward products (some of them follow easily from similar known theorems for forward products) are given in [7]. One simple sufficient condition for the ergodicity of backward products is that the series

$$\sum_{n=1}^{\infty} \min_{i,j} p_{ij}^{(n)}, \qquad P_n = (p_{ij}^{(n)}),$$

diverges (see [7]). A number of recent results on ergodicity of forward products is given in many books; see, for example, Seneta's book [37]. It is clear from Theorem 3.1 that when all $P_n \equiv P$, then strong ergodicity for the products P^n is equivalent to their weak ergodicity. Now a classical Markov theorem (proven by S. N. Bernstein in 1946) says that the forward products $P_{k,n}$ are weakly ergodic (for all k as $n \to \infty$) if the series $\sum_{n=1}^{\infty} \max_j \{\min_i p_{ij}^{(n)}\}$ diverges. It follows that when all $P_n \equiv P$, the products P^n is strongly ergodic if P has a column where all the entires are positive. In fact, if P is Markov, i.e., if P^k (for some k) has a column with all entries positive, then

$$\lim_{n \to \infty} P^{kn} = Q,$$

where Q is a stochastic matrix with identical rows. Notice that $P^s Q = Q$ for

---

[3] Backward products are introduced here not only for their appearance in the consensus problem, but also to demonstrate that a key multiplication property makes a difference in the ergodic behavior between these and forward products.

all positive integers s. This means that
$$\lim_{n\to\infty} P^{kn+s} = Q, \quad 0 \leq s \leq k.$$
In other words, any Markov matrix is strongly ergodic. The converse, namely, that a stochastic matrix P, if ergodic, is Markov, is trivially true. We now describe briefly the rate of convergence of P^n to $Q = \lim_n P^n$. The rate of convergence of P^n to Q (with identical rows) is geometric. This can be shown by the following simple argument: Write[4]
$$\alpha(P) = \tfrac{1}{2} \max_{i,j} \sum_{l=1}^{m} |p_{il} - p_{jl}|.$$
Suppose $\lim_{n\to\infty} P^n = Q$ and define for $B = (b_{ij})$ the usual norm $\|B\| = \sup_i \sum_{j=1}^{m} |b_{ij}|$. Now for all positive integers n, it is easy to see that
$$QP^n = Q \quad \text{and} \quad \alpha(P^n) \leq [\alpha(P)]^n.$$
Since $\alpha(Q) = 0$, there is a positive integer k such that $\alpha(P^k) = c < 1$. For positive integers n and s, let us denote
$$P^s - Q = (d_{ij}), \quad P^{nk} = (f_{ij}).$$
Noticing that $\sum_{j=1}^{m} d_{ij} = 0$ for all i, we have
$$\|P^{s+nk} - Q\| = \|(P^s - Q)P^{nk}\| = \sup_i \sum_{j=1}^{m} \left(\sum_{l=1}^{m} d_{il} f_{lj} \right)$$
$$= \sup_i \sum_{j=1}^{m} \left(\sum_{l=1}^{m} d_{il}[f_{lj} - f_{ij}] + f_{ij} \sum_{l=1}^{m} d_{il} \right)$$
$$\leq \sup_i \sum_{l=1}^{m} \left(|d_{il}| \sum_{j=1}^{m} |f_{lj} - f_{ij}| \right)$$
$$\leq 2\alpha(P^{kn})\|P^s - Q\| \leq 4c^n.$$
The geometric convergence of P^n to Q is now obvious.

Now following [21], one can also formulate a theorem for the rate of convergence of the backward products $P'_{k,n} = P_n P_{n-1} \cdots P_k$.

Theorem 3.2 *Suppose P is strongly ergodic and $\|P^n - Q\| < Ac^n, 0 < c < 1$, where Q is a stochastic matrix with identical rows. Let $g(n)$ be a monotonic increasing function from the positive integer into the positive real numbers*

---

[4] $\alpha(P)$ is called the ergodic coefficient of P and was first introduced by Dobrushinin *Theor. Probability Appl.* **1** (1956), 329–383. (See also [21] for discussion of the rate.)

such that

$$\lim_{n \to \infty} g(2n)\|P_n - P\| = 0.$$

Then the following result holds:

$$\lim_{n \to \infty} \sup_k \{\min(\lambda^n, g(n))\|P'_{k,k+n} - Q\|\} = 0,$$

where $1 < \lambda < \sqrt{1/c}$.

We omit the proof. The proof is similar to that for forward products given in [21]. So far, we have given above a very brief description of the ergodicity problem for products of stochastic matrices, and shown some specimens of known results that are available in this area. Now we are going to show a connection between ergodic theorems for Markov chains and limit theorems for convolution products of probability measures on semigroups. We believe that such a connection can lead sometimes to better (or new) insights into the ergodic problems for products of stochastic matrices.

First, we give a proof (the idea of the proof is not new) of the classical Birkhoff theorem [1] that a bistochastic $m \times m$ matrix $P = (P_{ij})$ (i.e., where $\sum_{j=1}^{m} p_{ij} = \sum_{i=1}^{m} p_{ij} = 1$ for each i and j, and $p_{ij} \geq 0$) can be expressed as a convex combination of permutation matrices (i.e., matrices that are obtained by permuting the columns of the $m \times m$ identity matrix). Note that the permutation matrices form a group of order $m!$.

Theorem 3.3 *Let P be an $m \times m$ stochastic matrix. Then P is a convex combination of the extreme stochastic matrices, i.e., stochastic matrices where the entries are all 0's and 1's. If P is bistochastic, then P is a convex combination of permutation matrices. These representations are not unique.*

Proof (There are many known ways to prove this theorem. But here we will apply the Krein–Milman theorem, though other elementary proofs can be found.) The set of all $m \times m$ matrices can be identified with R^{m^2} as a topological vector space. Then the set of all $m \times m$ stochastic matrices is a compact convex set. The extreme points of this compact set are clearly the extreme stochastic matrices. Hence by the Krein–Milman theorem [12, p. 440], P lies in the closed convex hull of the extreme points. Since the extreme points here are only finitely many, P is a convex combination of the extreme stochastic matrices. In the bistochastic case, the proof that the extreme points are precisely the permutation matrices is less simple. The permutation matrices are clearly extreme points. To prove the converse, let M be a bistochastic nonpermutation matrix. The proof is complete when we show that M is not an extreme point of the compact set of bistochastic matrices. Note that M has at least one entry which is neither zero nor one. Any row

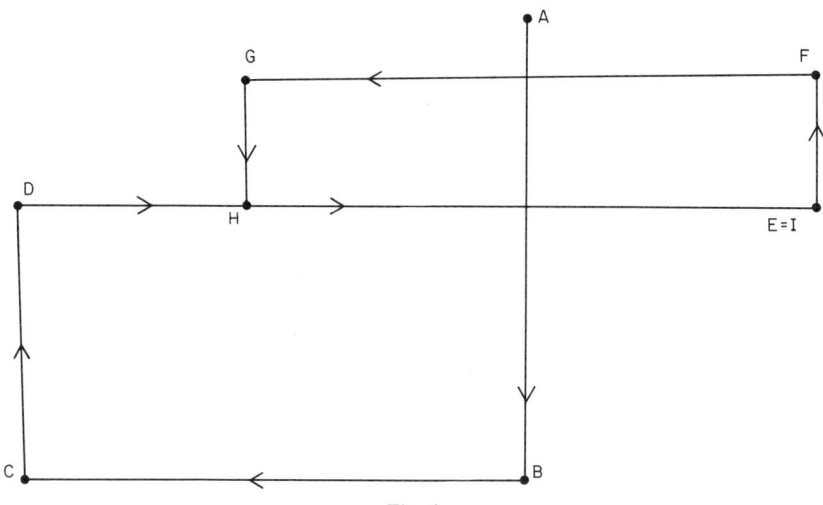

Fig. 4

or column of M having such an entry has at least two such entries. Examining any such matrix (see Fig. 4, where the dots represent entries $\neq 0$, $\neq 1$—this figure shows how one can get these dots until one is repeated), it is easy to see that one can always find i rows and i columns ($2 \leq i \leq m$) such that the elements which appear in these row column positions (G, H, E, F in Fig. 4) are *all* different from both 0 and 1; in each of these rows and columns, there are exactly two such row–column positions. Let $\varepsilon > 0$ be smaller than all the entries in these positions. Let M_ε be the matrix which is obtained from M by changing the entries of M *only* at the above row–column positions by adding ε and subtracting ε and from the entries in the way shown in Fig. 5.

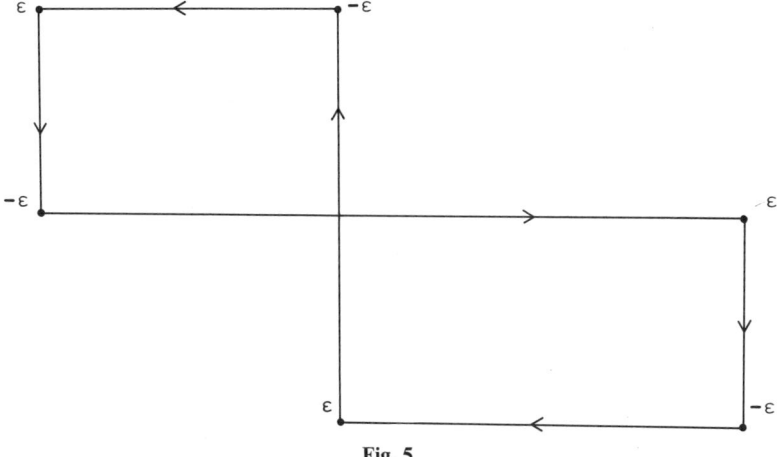

Fig. 5

Then M_ε is a bistochastic matrix different from M. Obtain $M_{-\varepsilon}$ similarly, so that

$$M = \tfrac{1}{2}M_\varepsilon + \tfrac{1}{2}M_{-\varepsilon}.$$

This proves that M is not an extreme point. Q.E.D.

For different proofs of the above theorem in the bistochastic case, the reader can consult Birkhoff [1] and Mirsky [31]. In the stochastic case, we show a more direct and elementary way to prove the theorem as follows: Let $P = (p_{ij})$ be a stochastic matrix. Then we can verify easily that

$$P = \sum_{\substack{k_1, k_2, \ldots, k_m \\ 1 \leq k_i \leq m}} p_{1k_1} p_{2k_2} \cdots p_{mk_m} A_{k_1 k_2 \cdots k_m}, \qquad (3.2)$$

where $A_{k_1 k_2 \cdots k_m}$ is the stochastic matrix (a_{ij}) such that

$$a_{ij} = \begin{cases} 1 & \text{if } j = k_i \\ 0 & \text{otherwise.} \end{cases}$$

Let us now denote the set of all extreme $m \times m$ stochastic matrices by S. Notice that S is a finite semigroup of order m^m. Let P_1 and P_2 be any two stochastic matrices. Then by Theorem 3.3, there exist nonnegative numbers (a_i) and (b_i) such that $\sum_{i=1}^{m^m} a_i = 1$, $\sum_{i=1}^{m^m} b_i = 1$, and

$$P_1 = \sum_i a_i A_i, \qquad P_2 = \sum_i b_i A_i,$$

where $S = \{A_1, A_2, \ldots, A_{m^m}\}$. We now associate P_1 with the probability measure μ_1 defined by $\mu_1(A_i) = a_i$. Similarly, we associate P_2 with a probability measure μ_2 on S. Notice that $\mu_1 \mu_2$ is the measure given by

$$\mu_1 \mu_2(A_i) = \sum_{A_k A_j = A_i} a_k b_j \qquad (= c_i, \text{ say}),$$

and $P_1 P_2 = \sum_i c_i A_i$. In other words, though the representation of stochastic matrices as convex combinations of extreme stochastic matrices is not unique, the measure $\mu_1 \mu_2$ is always given by one of the representations of $P_1 P_2$. Also it follows that if $P_1, P_2, \ldots$ is a sequence of $m \times m$ stochastic matrices and $\mu_1, \mu_2, \ldots$ the corresponding sequence of probability measures on S obtained as above, then $P_1 P_2 \cdots P_n$ converges as $n \to \infty$ whenever $\mu_1 \mu_2 \cdots \mu_n$ converges weakly (i.e., pointwise in S in this case). We also observe that if P is an $m \times m$ stochastic matrix with ith column entries all positive, then we can always find a representation of P via (3.2) such that the corresponding probability measure μ will have positive mass at that element of S where all the 1's appear in the ith column. We can make a similar remark for bistochastic matrices, which, however, needs a small justification: Suppose π is a permutation of $\{1, 2, \ldots, m\}$ into itself and $P =$

(p_{ij}) is a bistochastic matrix so that for each i, $1 \leq i \leq m$, $p_{i\pi(i)} > 0$. Choose $d > 0$ such that $d < p_{i\pi(i)}$, $1 \leq i \leq m$. If $D = (d_{ij})$ is the permutation matrix such that for each i, $d_{i\pi(i)} = 1$ then, $[1/(1-d)][P - d \cdot D]$ is a bistochastic matrix. By Theorem 3.3, there exist nonnegative numbers a_i such that $\sum_i a_i = 1$ and permutation matrices P_i such that

$$\frac{1}{1-d}[P - d \cdot D] = \sum_i a_i P_i.$$

This means that

$$P = dD + \sum_i a_i(1-d)P_i$$

so that we can correspond P to the probability measure μ, where $\mu(D) = d > 0$.

Now we state a result from [32] describing a convergence theorem for the iterates of a probability measure. We will apply this result here to obtain a completely different proof of the well-known ergodicity theorem that we discussed earlier in this section.

Theorem 3.4 *Suppose μ is a probability measure on a compact topological semigroup (or a completely simple semigroup with a compact group factor). Let F be the support of μ. If k is the smallest positive integer such that $F^n \cap F^{n+k} \neq \emptyset$ for some positive integer n, then for each positive integer s with $0 \leq s \leq k$, the sequence μ^{s+nk} converges weakly as $n \to \infty$.*

Now we use this theorem to prove

Theorem 3.5 *Suppose P is an $m \times m$ stochastic matrix such that for some positive integer k, P^k has one of its columns entirely positive. Then P is ergodic, i.e., $\lim_{n \to \infty} P^n = Q$, where Q is a stochastic matrix with identical rows. Furthermore, if P is ergodic, then P^k, for some positive integer k, has one of its columns entirely positive.*

Proof Let $P_1 = P^k$ and μ be the corresponding probability measure on S, the semigroup of extreme stochastic matrices. Let P_1 have its first column entirely positive. Let e_1 be the stochastic matrix where all the 1's appear in the first column. We can, of course, choose μ so that $\mu(e_1) > 0$. Let F be the support of μ. Let $H \subset S$ be the semigroup generated by F. Since $e_1 \in H$, all the elements of H that have rank 1 form the smallest two-sided ideal K of H. It is a well-known fact (see [34, p. 33]) that every weak limit point of the sequence (μ^n) has its support contained in K. By Theorem 3.4 plus the fact that S is a finite semigroup, it follows that for some positive integer s, μ^{ns} converges weakly as $n \to \infty$ to a probability measure whose support is contained in K. Then $P_1^{ns} = P^{nks}$ converges to a matrix Q with identical rows. Since $P^r Q = Q$ for all positive integers r, it follows that

$$P^{r+nks} \to Q, \qquad 0 \leqslant r < ks.$$

Hence, $\lim_{n \to \infty} P^n = Q$. The converse is trivial. Q.E.D.

We may now remark that because of Theorem 3.4, for any $m \times m$ stochastic matrix P, there is a positive integer s ($\geqslant 1$) such that for $0 \leqslant r < s$, $\lim_{n \to \infty} P^{ns+r} = Q_r$ exists. Using Theorem 3.4, we can also prove the following theorem.

Theorem 3.6 *Suppose that $P = (p_{ij})$ is an $m \times m$ stochastic matrix. Suppose that for each j, $1 \leqslant j \leqslant m$, there exists i_j ($1 \leqslant i_j \leqslant m$ and i_j possibly equal to j) such that*

$$p_{ji_j} > 0$$

and

$$p_{i_j i_j} > 0;$$

then $\lim_{n \to \infty} P^n$ exists.

Notice that if P has a positive diagonal, then P has this property. This property is, however, far more general than having a positive diagonal.

Proof Notice that an extreme stochastic matrix is idempotent if and only if it has the property described in the theorem. Hence, the probability measure μ corresponding to the stochastic matrix P (under a suitable representation) will have an idempotent element in its support F. Consequently, $F \cap F^2 \neq \emptyset$. By Theorem 3.4, the sequence μ^n converges weakly as $n \to \infty$ so that $\lim_{n \to \infty} P^n$ also exists. Q.E.D.

Note that in the above theorem, the limit matrix need not have its rows identical. It is also possible to make certain nonobvious remarks on limit theorems for matrix products which follow almost immediately from certain limit theorems for probability measures. For example, it is well known (see [34]) that an idempotent probability measure on a group is merely the uniform distribution on a subgroup. Now if B is a bistochastic idempotent matrix, we can consider a probability measure μ on the group of permutation matrices (via the Birkhoff theorem). Though μ^n, for each positive integer n, is given by one of the representations of B^n ($=B$, here) as convex combinations of permutation matrices, it is not clear whether μ has to be an idempotent probability measure. However, from Theorem 3.4, we know that on a finite semigroup, there is always a positive integer d such that μ^{nd} converges weakly as $n \to \infty$ to an idempotent probability measure λ. From the properties of correspondence, it now follows that B^{nd} ($=B$) converges to the matrix corresponding to λ. Since λ is idempotent, it is clear that B must be of the form

$$B = \frac{1}{s} \sum_{i=1}^{s} A_i,$$

where $\{A_1, A_2, \ldots, A_s\}$ is a subgroup of the permutation group. The converse is, obviously, true. Note that the correspondence between bistochastic idempotent matrices and idempotent probability measures on the permutation group is not one-one. For example, the matrix

$$\begin{bmatrix} \frac{1}{3} & \frac{1}{3} & \frac{1}{3} \\ \frac{1}{3} & \frac{1}{3} & \frac{1}{3} \\ \frac{1}{3} & \frac{1}{3} & \frac{1}{3} \end{bmatrix}$$

corresponds to the uniform distribution on the subgroup $\{e, (123), (132)\}$ and also to the uniform distribution on the entire permutation group. We now make a remark on the convergence of powers of a bistochastic matrix. Suppose that A is any bistochastic matrix such that

$$A = \sum_{i=1}^{p} a_i A_i,$$

where $a_i \geq 0$, $\sum_i a_i = 1$, and the A_i's are permutation matrices. Then if there is a word from the alphabet $\{A_1, A_2, \ldots, A_p\}$ of length n (repetitions permitted) which can also be expressed as a word of length $n+1$ for some positive integer n, then $\lim_{n \to \infty} A^n$ exists and is a bistochastic idempotent matrix. We can also make certain assertions about the limits of non-homogeneous convergent bistochastic chains. One such assertion is given in the following theorem, where, as we will see in another section, a general semigroup idea is involved.

Theorem 3.7 *Let (P_n) be a sequence of $m \times m$ bistochastic matrices such that for each positive integer k, the sequence $P_{k,n}$ converges elementwise to the matrix Q_k. Then $\lim_{n \to \infty} Q_k = Q_\infty$ and Q_∞ is a bistochastic idempotent matrix.*

Proof Let Q' be a limit point of the Q_k's. Now for $k < l < n$, $P_{kl} P_{ln} = P_{kn}$, so that by passing n to infinity we have

$$P_{kl} Q_l = Q_k, \quad 1 \leq k < l < \infty.$$

Therefore, it follows that

$$Q_k Q' = Q_k, \quad 1 \leq k < \infty.$$

If Q'' is another limit point of the Q_k's, then we must have

$$Q'' Q' = Q''$$

In particular, $Q'Q' = Q'$. Now we observe that if λ_1, λ_2 are uniform distributions on subgroups of permutation matrices and $\lambda_1\lambda_2 = \lambda_1, \lambda_2\lambda_1 = \lambda_2$, then their supports are identical, and consequently, $\lambda_1 = \lambda_2$. This implies that $Q' = Q''$.[5] Q.E.D.

Note that the natural setting for the existence of results like above is the class of topological semigroups. The above result holds in any compact topological semigroup (where the topology is first countable) and where for idempotents e_1 and e_2, $e_1e_2 = e_1$ and $e_2e_1 = e_2$ hold only when $e_1 = e_2$.

Consideration of semigroup ideas often gives us better insights into seemingly unrelated areas of study. For instance, if we go back to the discussion of idempotent matrices, we can make several non-obvious assertions about general stochastic idempotent matrices. Of course by Theorem 3.5, if a stochastic idempotent matrix has one of its columns entirely positive, then it must have all its rows identical. It is clear from the proof of Theorem 3.5 that any stochastic idempotent matrix is always a convex combination of 0–1 stochastic matrices of the same rank.

In what follows, we will discuss another way of corresponding probability measures on semigroups to stochastic matrices. This was discussed briefly by Martin-Löf in [30] and then used to study homogeneous stochastic processes on finite semigroups via the theory of Markov chain. This correspondence was also used in [24].

Let S be a finite semigroup. If S has no identity, then we adjoin an identity e to S and define multiplication in $S \cup \{e\}$ as follows:

$$xe = ex = x \quad \text{for } x \in S \quad \text{and} \quad ee = e.$$

To study limit theorems for probability measures on S, it is no loss of generality to assume that S has an identity. Suppose that m is the number of elements of S (counting the identity, adjoined if necessary). We now define an isomorphism from S into the semigroup of stochastic matrices by

$$\Phi(s) = A_s, \quad s \in S,$$

where

$$S = \{a_1, a_2, \ldots, a_m\}$$

and A_s is defined by

$$(A_s)_{i,j} = \begin{cases} 1 & \text{if } a_is = a_j \\ 0 & \text{otherwise.} \end{cases}$$

[5] Choose a uniform measure λ_1 (λ_2) for Q' (Q'') with the largest support. Then $\lim_{n\to\infty}(\lambda_1\lambda_2)^n = \mu = \mu^2$ exists and $\mu\lambda_1\lambda_2 = \mu$. Since $S_{\lambda_1} \subset S_\mu$, $\lambda_1 = \mu = \lambda_1\lambda_2$. Similarly, $\lambda_2\lambda_1 = \lambda_2$. In the stochastic case, $Q''Q' = Q''$ also holds, but Q' may not equal Q''.

It is easily verified that Φ is one-one (this is where we need an identity) and $\Phi(a_i a_j) = \Phi(a_i)\Phi(a_j)$. For $\mu \in P(S)$, we define

$$P_\mu = \sum_{i=1}^{m} p_i A_{a_i}, \quad \mu(a_i) = p_i.$$

Notice that if $a_i^{-1} a_j \equiv \{x \in S : a_i x = a_j\}$, then

$$P_\mu = (p_{ij}), \quad p_{ij} = \mu(a_i^{-1} a_j).$$

Again it can be verified that the correspondence

$$\mu \to P_\mu$$

is a homomorphism from $P(S)$ (with convolution as multiplication) into the semigroup of stochastic matrices.

Let (μ_n) be a sequence in $P(S)$. Notice that if the sequence of matrices P_{μ_n} is weakly ergodic, then for each positive integer k and for all i, j ($1 \leq i, j < m$), we have

$$|\mu_{k,n}(a_j) - \mu_{k,n}(a_i^{-1} a_j)| \to 0$$

as n tends to infinity. Here, $\mu_{k,n}$ denotes, of course, the convolution product $\mu_{k+1} \mu_{k+2} \cdots \mu_n$.

Notice that if S is a group, then for each j,

$$\{a_i^{-1} a_j \mid 1 \leq i \leq m\} = S$$

and therefore, for $1 \leq j, 1 \leq m$,

$$\lim_{n \to \infty} |\mu_{k,n}(a_j) - \mu_{k,n}(a_1)| = 0;$$

this implies that $\lim_{n \to \infty} \mu_{k,n}(a_j) = 1/m$.

Now it is easy to state a theorem on convolution products of probability measures on finite groups. This follows immediately from an ergodicity result of Wolfowitz [40], extended by Seneta, the extension being stated as follows: Suppose that there is a sequence of $m \times m$ stochastic matrices $\{P_1, P_2, \ldots\}$ such that for some $\delta > 0$ and all positive integers n,

$$p_{ij}^{(n)} \in \{0\} \cup [\delta, 1], \quad P_n = (p_{ij}^{(n)}).$$

Also suppose that any word (of finite length, repetitions allowed and counted) in the P's is ergodic. Then given $\varepsilon > 0$, there is a positive integer N such that for any word W in the P's of length $n > N$,

$$\max_{i,j} \sum_{l=1}^{m} |W_{il} - W_{jl}| < \varepsilon.$$

Theorem 3.8 Let $S \equiv \{a_1, a_2, \ldots, a_m\}$ be a finite group of order m. Let (λ_n) be a sequence in $P(S)$ with supports (S_{λ_n}) satisfying the following conditions:

(i) $e \in S_{\lambda_n}$ for each n;
(ii) there is a positive number δ such that for all n,
$$\lambda_n(a_i) \in \{0\} \cup [\delta, 1], \quad 1 \leqslant i \leqslant m;$$
(iii) for each n, there is a positive integer k_n such that
$$S = S_{\lambda_n}^{k_n}.$$

Then given $\varepsilon > 0$, there is a positive integer N such that for all mappings π of the positive integers into itself, we have for $1 \leqslant i \leqslant m$ and $n > N$:
$$\left|\lambda_{\pi(1)}\lambda_{\pi(2)} \cdots \lambda_{\pi(n)}(a_i) - (1/m)\right| < \varepsilon.$$

Proof Suppose P_n is the stochastic matrix corresponding to λ_n via the homomorphism Φ as described earlier. Let W be any word in the P's containing the letter P_1. Then for $1 \leqslant i, j \leqslant m$, we have
$$(W^{k_1})_{ij} > 0,$$
since the support of the measure corresponding to the matrix W^{k_1} contains $S_{\lambda_1}^{k_1} (\equiv S)$. By Theorem 3.5 and Seneta's result, the theorem follows. Q.E.D.

Before we close this section, we will present briefly some results on the convergence of nonhomogeneous bistochastic Markov chains. Here again we will use some results on the limit theory for convolution products of probability measures on groups. Two relevant theorems in this context are the following.

Theorem 3.9 Let S be an at most countable discrete group and (μ_n) be a sequence of probability measures on S. Then for all nonnegative integers k, the sequence $\mu_{k,n} = \mu_{k+1}\mu_{k+2} \cdots \mu_n$ converges weakly to some probability measure if and only if there exists a finite subgroup G such that the series $\sum_{n=1}^{\infty} \mu_n(S - G)$ converges, and for any proper subgroup G' of G and any choice of elements g_n in S, the series
$$\sum_{n=1}^{\infty} \mu_n(S - g_{n-1}G'g_n^{-1}) = \infty.$$

Theorem 3.10 Let the μ_n's and S be as in the previous theorem. Then a sufficient condition for the weak convergence of the sequence $\mu_{k,n}$ (for all nonnegative integers k) is that (i) there exists a finite subgroup G such that $\sum_{n=1}^{\infty} \mu_n(S - G)$ is convergent and (ii) $\mu_n(e) > s > 0$ for all n, where e is the identity of S. Also in case of convergence, $\lim_{k \to \infty} \lim_{n \to \infty} \mu_{k,n}$ exists and is an idempotent probability measure.

We omit the proofs of Theorem 3.9 and 3.10. These results are due to Center and Mukherjea [6]. An earlier finite group version of Theorem 3.10 was obtained first by Maksimov in [27] (see also Heyer [18]).

Actually, Theorem 3.10 is a corollary of Theorem 3.9 and can be obtained easily. Theorem 3.10 has an immediate application in the convergence theory for bistochastic chains.

Theorem 3.11 *Let $P_n \equiv (p_{ij}^{(n)})$ be a sequence of $m \times m$ bistochastic matrices such that for some $\delta > 0$, $p_{ii}^{(n)} \geqslant \delta$ for all $n \geqslant N$ (some positive integer) and all $i = 1, 2, \ldots, m$. Then for all k, the sequence $P_{k,n}$ is convergent.*

Proof Let μ_n be the probability measure on the group of $m \times m$ permutation matrices corresponding to the matrix P_n (via the Birkhoff theorem). Note that we can and do choose the μ_n's so that for $n \geqslant N$, $\mu_n(e) \geqslant \delta$. By Theorem 3.10, the sequence $\mu_{k,n}$ converges weakly. It follows that $P_{k,n}$ is convergent. Q.E.D.

From remarks prior to Theorem 3.7, we have seen that if B is a bistochastic $m \times m$ idempotent matrix, then B corresponds to the uniform distribution on a subgroup of permutation matrices. Each such subgroup uniquely determines a partition of the set $\{1, 2, \ldots, m\}$ in the sense that for any element π in the subgroup and for $1 \leqslant i \leqslant m$, i and $\pi(i)$ both belong to the same member of the partition. Also such a partition uniquely determines the bistochastic idempotent matrix. Thus if

$$\{1, 2\}, \quad \{3, 4\}$$

partition $\{1, 2, 3, 4\}$, then this partition determines the matrix

$$\begin{bmatrix} \frac{1}{2} & \frac{1}{2} & 0 & 0 \\ \frac{1}{2} & \frac{1}{2} & 0 & 0 \\ 0 & 0 & \frac{1}{2} & \frac{1}{2} \\ 0 & 0 & \frac{1}{2} & \frac{1}{2} \end{bmatrix}$$

which is idempotent. Notice that if

$$C_1, C_2, \ldots, C_p$$

is a partition of $\{1, 2, \ldots, m\}$, then the corresponding bistochastic idempotent matrix B is given by

$$B = (b_{ij}), \quad b_{ij} = \begin{cases} 1/c_k & \text{if } i, j \in C_k \\ 0 & \text{if } i, j \text{ belong to different } C\text{'s,} \end{cases}$$

where c_k is the number of elements in C_k.

Now we need a definition.

Definition 3.3 Let (P_n) be a sequence of bistochastic $m \times m$ matrices such that $\lim_{n \to \infty} P_{k,n} = B_k$ exists. Then $\lim_{k \to \infty} B_k = B$ exists (see Theorem 3.7) and $B = B^2$. The partition of $\{1, 2, \ldots, m\}$ corresponding to B is called the *basis* of the convergent bistochastic chain (P_n).

Now we will prove a theorem of Maksimov on convergent bistochastic chains as another application of Theorem 3.10. To prove the theorem, we will utilize the following lemma. The lemma in the following form is not given in [28]. The form in which it appears below reveals an interesting interplay between bistochastic matrices and corresponding probability measures. We use a general semigroup idea to prove the lemma.

Lemma 3.1 *Let $P_n = (p_{ij}^{(n)})$ be a convergent bistochastic chain, that is, $\lim_{n \to \infty} P_{k,n} = B_k$ exists for each nonnegative integer k. Let $B = \lim_{k \to \infty} B_k$ (see Theorem 3.7). Suppose that*

$$\{C_1, C_2, \ldots, C_p\}: \text{a partition of } \{1, 2, \ldots, m\}$$

is the basis of the chain. For each n, let μ_n be the probability measure on the group of permutation matrices corresponding to P_n. Then for any sequence (n_i) of positive integers, there is a subsequence (n_{i_j}) such that for each positive integer k,

$$\mu_{k, n_{i_j}} \to \pi_k \in P(G)$$

and $\pi_{n_{i_j}} \to \pi_\infty = \pi_\infty^2 \in P(G)$ as j tends to infinity. Here π_∞ is an idempotent probability measure (and therefore, a uniform distribution on a subgroup G' of G) corresponding to B, and therefore, for g in G', $g(i) = j$ if and only if both i and j belong to the same C_k. Furthermore, if for some g in G,

$$\mu_n(g) \geqslant \varepsilon > 0$$

for infinitely many n, then also $g(i) = j$ only if both i and j belong to the same C_k.

Proof The first part of the lemma follows from the general compact semigroup theorem given in Section 1. Now

$$\mu_{k, n_{i_j}} \to \pi_k, \qquad \pi_{n_{i_j}} \to \pi_\infty$$

imply that π_k corresponds to the matrix B_k and π_∞ corresponds to the matrix B. Since the support of any uniform distribution, which corresponds to the bistochastic matrix B, gives rise to the same partition of $\{1, 2, \ldots, m\}$ characterizing the matrix B, every permutation in the support of π_∞ permutes elements within the same members of the partition $\{C_i\}$.

To prove the last part of the lemma, suppose that for the subsequence (n_i) of positive integers,

$$\mu_{n_i + 1}(g) > \varepsilon > 0, \qquad i \geqslant 1.$$

Then by the first part of the lemma, we can find a suitable subsequence (n_{i_j}) of (n_i) such that

$$\mu_{k,n_{i_j}} \to \pi_k, \qquad \mu_{k,n_{i_j}+1} \to \pi'_k,$$
$$\pi_{n_{i_j}} \to \pi_\infty, \qquad \pi'_{n_{i_j}+1} \to \pi'_\infty,$$

where π_∞ and π'_∞ are idempotent probability measures on G, corresponding to the matrix B. Now we observe that

$$\mu_{k,n_{i_j}+1}(g) \geqslant \mu_{k,n_{i_j}}(e)\mu_{n_{i_j}+1}(g) \geqslant \varepsilon \mu_{k,n_{i_j}}(e).$$

Since e belongs to the support of π_∞, it follows that g is in the support of π'_∞. The lemma now follows from earlier remarks.   Q.E.D.

The following lemma is also needed.

Lemma 3.2 *Let G be a finite group and μ_n be a sequence of probability measures on G such that for each positive integer k, $\lim_{n\to\infty} \mu_{k,n} = \pi_k$ exists. If ω_H (the uniform distribution on a subgroup H) is a limit point of $\{\pi_k\}$, then*

$$\sum_{n=1}^\infty \mu_n(G - H) < \infty.$$

Proof Choose positive integers p and n_0 such that for all $n > n_0$, we have $\mu_{p,n}(H) > 2/3$. Then after simple computations,

$$\mu_{p,n+1}(H) \leqslant \mu_{p,n}(H) - \tfrac{1}{3}[1 - \mu_{n+1}(H)].$$

Repeating the same process, we have

$$\mu_{p,n+s}(H) \leqslant \mu_{p,n}(H) - \tfrac{1}{3} \sum_{i=n+1}^{n+s} \mu_i(G - H),$$

The lemma is now clear.   Q.E.D.

Now we are ready to prove Maksimov's theorem.

Theorem 3.12 *Let $P_n = (p_{ij}^{(n)})$ be a convergent bistochastic chain so that $\lim_{k\to\infty} \lim_{n\to\infty} P_{k,n} = B$. Let $\{C_1, C_2, \ldots, C_p\}$ be the basis of the chain. Then the series*

$$\sum_{n=1}^\infty p_{ij}^{(n)} < \infty$$

whenever i, j belong to different members of the partition $\{C_i\}$.

Proof This proof (based mainly on Maksimov's ideas) is not generalizable to stochastic chains. For a new proof that is generalizable, see Note 3 added in proof at end of paper. Let μ_n be a probability measure on the group G of permutation matrices corresponding to P_n. Choose $g_n \in G$ such that for each n, $\mu_n(g_n) \geqslant 1/m!$.

Since there are only $m!$ elements in G, there is a positive integer k_0 such that for any $k > k_0$, g_k equals g_n for infinitely many n. Choose $k > k_0$ and consider the measure

$$\lambda_n = g_k g_{k+1} \cdots g_{n-1} \mu_n (g_k g_{k+1} \cdots g_n)^{-1}.$$

Then for each n,

$$\lambda_n(e) = \mu_n(g_n) \geq 1/m!.$$

By Theorem 3.10, we have

$$\lim_{n \to \infty} \lim_{N \to \infty} \lambda_{n,N} = \lambda_\infty \quad (= \lambda_\infty^2),$$

where λ_∞ is the uniform distribution on a subgroup H. This means that

$$\lim_{n \to \infty} \lim_{N \to \infty} g_k g_{k+1} \cdots g_n \mu_{n,N} (g_k g_{k+1} \cdots g_N)^{-1} = \lambda_\infty.$$

Let Q be the bistochastic matrix corresponding to λ_∞. Then for some h_1, h_2 in G, we have

$$h_1 B h_2 = Q$$

since we have the following equality:

$$\lim_{n \to \infty} \lim_{N \to \infty} g_k g_{k+1} \cdots g_n P_{n,N} (g_k g_{k+1} \cdots g_N)^{-1} = Q.$$

By Lemma 3.1, the g_k's (and therefore, h_1 and h_2) permute elements within the same members of $\{C_1\}$ because of our choice of $k(> k_0)$. By Lemma 3.2, we have

$$\sum_{n=1}^\infty \lambda_n(G - H) < \infty, \quad H = \text{the support of } \lambda_\infty.$$

Notice that the elements of H permute elements within the same members of the partition $\{C_l\}$. Now let i and j belong to different members of $\{C_l\}$. Then for each n, only for $h_n \in G - H$ can we have

$$(g_k g_{k+1} \cdots g_{n-1})^{-1} h_n (g_k g_{k+1} \cdots g_n)(i) = j.$$

Since μ_n corresponds to P_n, it is clear that

$$p_{ij}^{(n)} \leq \lambda_n(G - H).$$

The theorem now follows. Q.E.D.

Two nonhomogeneous Markov chains will be called equivalent if

$$\sum_{n=1}^\infty |p_{ij}^{(n)} - p_{ij}'^{(n)}| < \infty$$

for $1 \leqslant i, j \leqslant m$. Note that it is not difficult to see from the inequality below that two Markov chains (P_n) and (P'_n), which are equivalent, are either both convergent or both divergent. The reason is the following:

Defining $\|P\| = \sup_i \sum_j |p_{ij}|$, we see that

$$\|P_k P_{k+1} \cdots P_n - P'_k P'_{k+1} \cdots P'_n\|$$
$$= \left\| \sum_{l=0}^{n-k} P_k P_{k+1} \cdots P_{k+l-1}(P_{k+l} - P'_{k+l}) P'_{k+l+1} \cdots P'_n \right\|^6$$
$$\leqslant \sum_{l=0}^{n-k} \|P_{k+l} - P'_{k+l}\|$$

$\to 0$ as $k \to \infty$, if the chains are equivalent. Two nonhomogeneous equivalent bistochastic chains, when convergent, have the same basis. This is another assertion made by Maksimov in [28]; this, of course, follows immediately from the above inequality. (See Note 2 added in proof at end of paper.)

We now close this section with the following remark. It is quite clear from the demonstration of various results in this section that the theory of probability measures on finite semigroups can help us better understand the different types of ergodic behavior of homogeneous as well as nonhomogeneous Markov chains. The theory of convergence of convolution products of nonidentical probability measures, though quite developed and somewhat complete in finite and at most countable groups (see [6, 27]), has not been looked into carefully and systematically in finite semigroups. When this is done, this will undoubtedly enrich and add to the already existing large literature on general ergodic Markov chains. We also suggest the possibility of studying the ergodic problems for denumerable Markov chains (especially in the bistochastic case) by appealing to the various weak convergence results for convolution products of probability measures on countable groups via the extension of the Birkhoff theorem to the case of infinite doubly stochastic matrices. Indeed, this extension has been carefully studied during 1955–1965 by various authors, including D. G. Kendall, J. Kiefer, J. R. Isbell, P. Révész, and many others. We will briefly describe this extension in the appendix to this section.

Appendix

Our proof of the Birkhoff theorem is somewhat similar to that of J. von Neumann (who did not use the Krein–Milman theorem) in his article: "A certain zero-sum two person game equivalent to the optimal assignment

---

[6] The first factor in this product is omitted for $l = 0$; the last factor is omitted for $l = n - k$.

problem [*Ann. Math. Stud.* 2, No. 28 (1953), 5–12]. G. Birkhoff in problem 111 of his book "Lattice Theory" asked for an extension of his theorem to the case of countably infinite doubly stochastic matrices. Rattray and Peck [*Trans. Roy. Soc. Canada* III, No. 3 (1955), 49], Isbell [*Proc. Amer. Math. Soc.* 6 (1955)] and Kendall [25] all independently (though somewhat differently extended Birkhoff's theorem.

Let $\mathfrak{X}$ be the real vector space of countably infinite real matrices $A = (a_{ij})$ such that

$$\|A\| \equiv \max\left\{\sup_i \sum_j |a_{ij}|, \sup_j \sum_i |a_{ij}|\right\} < \infty.$$

Then $\mathfrak{X}$ is a Banach space under the above norm. Isbell first showed that the closed (closure in the metric topology of $\mathfrak{X}$) convex hull of the permutation matrices is a *proper* subset of the set of all doubly stochastic matrices. Kendall considered the above "closure" with respect to the weak topology on $\mathfrak{X}$ generated by the following linear functionals (the smallest topology so that these functionals are continuous):

$$f_i(A) = \sum_j A_{ij}, \qquad f^j(A) = \sum_i A_{ij}, \qquad \text{and} \qquad f_{i,j}(A) = A_{ij},$$

where $i, j = 1, 2, 3, \ldots$. He proved that the set of all infinite doubly stochastic matrices is exactly the closed convex hull of the set of all infinite permutation matrices. It was also shown in his paper (by an argument due to J. C. Kiefer) that the extreme points of the set of infinite doubly stochastic matrices is precisely the set of all infinite permutation matrices. Later, Isbell [23] also established this result by an elementary argument. He also showed that a doubly stochastic matrix $A = (a_{ij})$ in which the a_{ij}'s take only finitely many distinct values is a convex combination of permutation matrices.

Révész [35] first noticed that we can correspond to a doubly stochastic $n \times n$ matrix a probability measure on the set of all $n \times n$ permutation matrices. He looked at Birkhoff's theorem from this angle and extended it to the countably infinite case as follows: Let $A = (a_{ij})$, $i, j = 1, 2, \ldots$, be a doubly stochastic matrix. Let Ω be the set of all permutation matrices. Let

$$\Omega_{k,l} = \{(b_{ij}) \in \Omega : b_{k,l} = 1\}.$$

Then we can define a σ-algebra Σ of subsets of Ω and a probability measure P on Σ such that $\Omega_{k,l} \in \Sigma$ and $P(\Omega_{k,l}) = a_{k,l}$, $k, l = 1, 2, \ldots$. Révész used his extension of Birkhoff's theorem and a couple of random ergodic theorems to deduce a strong law of large numbers for a class of nonhomogeneous chains.

Theorem (Révész) *Let $X_1, X_2, \ldots$ be a discrete Markov chain with state space $\{b_1, b_2, \ldots\}$ such that*

(i) $\sum_{i=1}^{\infty} b_i = 0$; $\sum_{i=1}^{\infty} |b_i| < \infty$,
(ii) *the matrices $P_n = (p_{ij}^{(n)})$, $p_{ij}^{(n)} = P[X_n = b_j | X_{n-1} = b_i]$ are doubly stochastic.*

Let us define the norm of a doubly stochastic matrix A by

$$\|A\| = \sup_{x \in H_1} (\|Ax\|/\|x\|),$$

where H_1 contains those points $x = (x_1, x_2, \ldots) \in l_2$ for which $\sum_{i=1}^{\infty} x_i = 0$, $\sum_{i=1}^{\infty} |x_i| < \infty$, and $\|x\|^2 = \sum_{i=1}^{\infty} x_i^2$. We assume that

$$\|P_n\| \leq 1 - (C/n^{1-\varepsilon})$$

where C is an arbitrary positive constant and $0 < \varepsilon \leq 1$. Then

$$P\left[\frac{1}{n}(X_1 + X_2 + \cdots + X_n) \to 0\right] = 1.$$

For details, the reader can consult Révész's notes: Seminar on random ergodic theory, Aarhus University, 1961. (See Note 3 added in proof at the end of paper for a new proof of Theorem 3.12.)

IV. Limit Theorems for Convolution Products of Probability Measures on Completely Simple Semigroups

In Section III, we have shown how one can utilize results on the weak convergence of convolution products of probability measures on a finite group to obtain results on the convergence of nonhomogeneous bistochastic chains. Naturally, similar weak convergence results on finite semigroups are expected to lead to a useful theory of convergence for nonhomogeneous stochastic chains. In [6], we used Csiszár–Tortrat's results (see Section I) to obtain two useful results in finite groups stated as Theorems 3.9 and 3.10 in this paper. It is therefore natural to ask whether it is possible to obtain analogs of Csiszár–Tortrat's results described in Section 1 for semigroups. The most important class of semigroups that arises in the context of probabilistic study is, as we have mentioned in Section I, the class of completely simple semigroups. There are at least two reasons which account for our special attention to these semigroups. One of them is the fact that the supports of idempotent probability measures on a general locally compact semigroup are such semigroups; another reason is that the set of all recurrent points of a random walk induced by a probability measure on many general semigroups including compact semigroups, when nonempty, is a completely simple subsemigroup.

One of the main purposes of this section is to demonstrate what kind of analogs of Csiszár–Tortrat's results can be expected in completely simple semigroups. We will also briefly consider some convergence results in general semigroups. Thus, three interesting results (Theorems 4.3, 4.4, and 4.6) are given in general semigroups. As to Theorem 4.1, we may remark that Csiszár's group arguments do not extend to the semigroup situation. However, some of his ideas, as will be shown here, can be carried over.

Theorem 4.1 *Let S ($\equiv X \times G \times Y$, the usual product representation) be a locally compact, second countable, completely simple semigroup such that X and Y are compact and G not necessarily compact. Let (Q_n) be a sequence in $P(S)$. Then either for some nonnegative integer k,*

$$Q_{k,n} \to 0 \quad \text{vaguely as} \quad n \to \infty,$$

or there exists a sequence $x_n \in S$ such that for every nonnegative integer k,

$$Q_{k,n} x_n \to Q'_k \in P(S)$$

as n tends to infinity.

Proof We use several steps to complete the proof. The proof does not use any new arguments. The proof consists of verifying certain arguments used in the proofs of various theorems in [32].

Step I For each nonnegative integer k, let us write

$$a_{kn}(K) = \sup\{Q_{k,n}(Kx^{-1}) : x \in S\}.$$

Notice that for all $n > k$,

$$a_{kn}(K) \geqslant a_{k(n+1)}(K).$$

So we can write

$$a_k(K) = \lim_{n \to \infty} a_{kn}(K) \quad \text{and} \quad a_k = \sup\{a_k(K) : K \text{ a compact subset of } S\}.$$

Then we claim that $a_k = 0$ or 1.

Proof of Step I Suppose that $0 < a_k < 1$. Then we choose c such that

$$a_k < c < 1, \quad 0 < c(1 + c)/2 < a_k.$$

Let D be an arbitrary compact set $\subset S$. Then there exists a positive integer m such that

$$\sup_{x \in S} Q_{k,m}(Dx^{-1}) < c - \varepsilon, \tag{4.1}$$

where $0 < \varepsilon < c/2$. There exists a compact set A such that
$$Q_{k,m}(S - A) < (c/2) - \varepsilon. \qquad (4.2)$$
Write $E = A^{-1}D$. Then it can be verified that E is compact. For $y \in S - Ex^{-1}$,
$$Dx^{-1}y^{-1} \cap A = \emptyset$$
and therefore,
$$Q_{k,m}(Dx^{-1}y^{-1}) < (c/2) - \varepsilon. \qquad (4.3)$$
Therefore, for $n > m$, we have (using (4.1))
$$Q_{k,n}(Dx^{-1}) = \int Q_{k,m}(Dx^{-1}y^{-1})Q_{m,n}(dy)$$
$$\leqslant \int_{Ex^{-1}} + \int_{S-Ex^{-1}}$$
$$< (c - \varepsilon)Q_{m,n}(Ex^{-1}) + \left(\frac{c}{2} - \varepsilon\right)[1 - Q_{m,n}(Ex^{-1})]$$
$$= (c/2) + (c/2)Q_{m,n}(Ex^{-1}). \qquad (4.4)$$
Now if B is a compact set such that
$$Q_{k,m}(B) > (a_k + c)/2c,$$
then
$$Q_{k,n}(BEx^{-1}) = \int Q_{k,m}(BEx^{-1}y^{-1})Q_{m,n}(dy)$$
$$\geqslant Q_{k,m}(B)Q_{m,n}(Ex^{-1});$$
This means that for all $n > N$ (some positive integer),
$$Q_{m,n}(Ex^{-1}) \leqslant \frac{2c}{a_k + c} \cdot Q_{k,n}((BE)x^{-1}) \leqslant \frac{2c}{a_k + c} \cdot \frac{a_k + c}{2} = c.$$
It follows from (4.4) that for $n > N$, $n > m$, and all x in S, we have
$$Q_{k,n}(Dx^{-1}) < (c/2) + (c^2/2) < a_k;$$
this means that for any compact set D,
$$a_k(D) \leqslant (c/2) + (c^2/2) < a_k,$$
which is a contradiction. Step I is proved.

Step II Suppose $Q_{k,n}$ does not converge vaguely to 0 as $n \to \infty$. Then there exists a subsequence (n_i) of positive integers such that Q_{k,n_i} has a vague cluster point in $P(S)$. This is because in this case, $a_k = 1$.

Then given $0 < \varepsilon < \delta$, there are compact sets K_1 and K_2 and a subsequence (n_i) of positive integers such that

$$Q_{k,n_i}(K_1) > \delta, \quad \limsup_{n \to \infty} \sup_x Q_{k,n}(K_2 x^{-1}) > 1 - \varepsilon.$$

Hence there exists i_0 such that for $i > i_0$ and elements x_{n_i} in S, we have

$$K_1 \cap K_2 x_{n_i}^{-1} \neq \emptyset \quad \text{or} \quad x_{n_i} \in K_1^{-1} K_2$$

so that

$$Q_{k,n_i}(K_2(K_1^{-1}K_2)^{-1}) > 1 - \varepsilon. \tag{4.5}$$

Since $K_2(K_1^{-1}K_2)^{-1}$ can be easily verified to be compact, the vague cluster points of the subsequence (Q_{k,n_i}) are probability measures.

Step III In this step, we will complete the proof of this theorem. Consider the subsequence (n_i) from Step II. Let us write

$$w_i = (Q_{0,n_i}, Q_{1,n_i}, \ldots, Q_{n_i-1,n_i}, 0, 0, \ldots).$$

Then (w_i) is a sequence in $B(S)$, which is vaguely compact. Therefore, there is a subsequence $(p_i) \subset (n_i)$ such that for each nonnegative integer k',

$$Q_{k',p_i} \to Q'_{k'} \in B(S).$$

By Step II,

$$Q'_k \in P(S). \tag{4.6}$$

For $k' \leq k$,

$$Q_{k',k} Q'_k = Q'_{k'} \tag{4.7}$$

and for $k' > k$,

$$Q_{k,k'} Q'_{k'} = Q'_k. \tag{4.8}$$

By (4.6), (4.7), and (4.8), we have for all positive integers k',

$$Q'_{k'} \in P(S). \tag{4.9}$$

(Notice that the assumption that the factor X and Y in the product representation of S are compact is crucial in the derivation of (4.8) and also in concluding that the set $K_2(K_1^{-1}K_2)^{-1}$ in (4.5) is compact.) Let (q_i) be a subsequence of (p_i) such that

$$Q'_{q_i} \to Q_\infty. \tag{4.10}$$

Then we have from (4.8) and (4.10),

$$Q'_k Q_\infty = Q_\infty, \quad Q_\infty Q_\infty = Q_\infty. \tag{4.11}$$

LIMIT THEOREMS

[Notice that compactness of X is needed to establish that Q_∞ is a probability measure; this is done by showing that the sequence Q'_{p_i} is uniformly tight, by using (4.8).]

Let us now write $S_{Q_\infty} = \bigcap_{n=1}^{\infty} G_n$, where G_n is a sequence of open sets with compact closure. Now there exists a subsequence $(s_i) \subset (q_i)$ such that

$$Q_{s_i, s_{i+1}} \to Q_\infty \quad \text{vaguely as} \quad i \to \infty$$

and for each $i \geq 1, j \geq 1$,

$$Q_{s_i, s_{i+j}}(G_i) > 1 - (1/i).$$

Let m be any positive integer such that

$$s_i < m \leq s_{i+1}.$$

Then there exists z_m in S such that

$$Q_{s_i, m}(G_i z_m^{-1}) \geq \int Q_{s_i, m}(G_i y^{-1}) Q_{m, s_{i+2}}(dy)$$
$$= Q_{s_i, s_{i+2}}(G_i) > 1 - (1/i).$$

Let $z \in S_{Q_\infty}$. Then we claim that for all nonnegative integers t, the sequence $Q_{t, n} z_n z$ converges vaguely to the probability measure $Q'_t z$ as $n \to \infty$.) To prove our claim, let Q be a vague cluster point of the sequence $Q_{t, n} z_n z$. Then there is a subsequence m_j of positive integers such that

$$Q_{t, m_j} z_{m_j} z \to Q \quad \text{as} \quad j \to \infty.$$

Now replace the sequence m_j by a suitable subsequence (and still calling this subsequence m_j) such that for some subsequence (s_{i_j}) of (s_i), we have

$$s_{i_j} < m_j \leq s_{i_j+1}.$$

The sequence $Q_{s_{i_j}, m_j} z_{m_j}$ has a vague cluster point Q'; this Q' belongs to $P(S)$ by the way the z_m's have been chosen. Also, it follows from the construction of the z_m's that $S_{Q'} \subset S_{Q_\infty}$. Then we can write

$$Q_\infty Q' z(B) = \int Q_\infty(B z^{-1} y^{-1}) Q'(dy) = Q_\infty(B z^{-1}) = Q_\infty z(B),$$

by a result in [34, p. 22]. Since we have

$$Q_{t, m_j} z_{m_j} z = Q_{t, s_{i_j}} Q_{s_{i_j}, m_j} z_{m_j} z,$$

it is clear from (4.11) and above that

$$Q = Q'_t Q' z = (Q'_t Q_\infty) Q' z = Q'_t (Q_\infty Q') z = Q'_t z.$$

The proof of the theorem is complete. Q.E.D.

The above proof holds in any semigroup where $A \times B^{-1}$ and $A^{-1} \times B$ are compact for compact A and B.

While the problem of finding when the sequence $Q_{k,n} = Q_{k+1} Q_{k+2} \cdots Q_n$ converges weakly is completely solved for at most countable discrete groups (see [6]), the problem is open for discrete semigroups. It is, however, easy to present a simple (and perhaps, not obvious) result for discrete semigroups containing an identity.

Theorem 4.2 *Let S be an at most countable discrete semigroup containing an identity e. Let Q_n be a sequence in $P(S)$ such that*

$$\sum_{n=1}^{\infty} Q_n(S - e) < \infty.$$

Then for every nonnegative integer k, the sequence $Q_{k,n}$ converges weakly to some probability measure Q'_k; also, the sequence Q'_k converges weakly as $k \to \infty$ to the unit mass at e.

Proof Following the same argument given in [6] or in [34, p. 76], we can easily show that for any subset A of S and nonnegative integers k, n, and m with $k + 1 \leq n$ and $m \geq 1$,

$$|Q_{k,n}(A) - Q_{k,n+m}(A)| \leq \sum_{i=n+1}^{n+m} Q_n(S - e).$$

This means that the sequence $Q_{k,n}$ converges vaguely. By a simple application of the Borel–Cantelli lemma, it follows that given $\varepsilon > 0$, there is a positive integer k_0 such that for $n > k \geq k_0$, we have

$$Q_{k,n}(e) > 1 - \varepsilon$$

and

$$\liminf_{k \to \infty \; n > k} Q_{k,n}(e) = 1.$$

Since S may be infinite, it is not immediately clear that the vague limit of the sequence $Q_{k,n}$ is a probability measure. To this end, the following observation is helpful. We claim that for $k < p$,

$$Q'_k = Q_{k,p} Q'_p,$$

where $Q'_k = \text{w*-lim}_{n \to \infty} Q_{k,n}$. Note that the verification of our claim is not immediate since convolution of measures in semigroups is not even separately continuous in the vague topology. To prove our claim, suppose that there exists x in S such that for $i \geq 1$,

$$|Q_{k,p} Q_{p,n_i}(x) - Q_{k,p} Q'_p(x)| > 3\varepsilon > 0$$

for some subsequence (n_i) of positive integers. Using the regularity of $Q_{k,p}$, we can find a finite set K such that for sufficiently large i,

$$|Q_{k,p}Q_{p,n_i}(x) - Q_{k,p}Q'_p(x)|$$

$$\leqslant \left|\sum_{y \in K} Q_{p,n_i}(y^{-1}x)Q_{k,p}(y) - Q_{k,p}Q_{p,n_i}(x)\right|$$

$$+ \left|\sum_{y \in K} Q_{p,n_i}(y^{-1}x)Q_{k,p}(y) - \sum_{y \in K} Q'_p(y^{-1}x)Q_{k,p}(y)\right|$$

$$+ \left|\sum_{y \in K} Q'_p(y^{-1}x)Q_{k,p}(y) - Q_{k,p}Q'_p(x)\right|$$

$$< \varepsilon + \varepsilon + \varepsilon = 3\varepsilon.$$

This is a contradiction. The claim is now verified and the rest of the proof is clear. Q.E.D.

An immediate corollary of the above theorem is that if (P_n) is a sequence of $m \times m$ stochastic matrices such that $P_n = (p_{ij}^{(n)})$ and

$$\sum_{n=1}^{\infty} \left[1 - \min_i p_{ii}^{(n)}\right] < \infty,$$

then for each nonnegative integer k, the sequence $P_{k,n} = P_{k+1}P_{k+2} \cdots P_n$ converges elementwise to a stochastic matrix. The proof follows easily following methods in Section III.

The problem of finding when the sequence Q^n of convolution iterates of a probability measure Q converges weakly or vaguely is completely solved in discrete and compact semigroups. For an account of these results, the reader can consult [30, 34, 36]. In these situations, the problem is relatively simple (compared to the general nondiscrete or noncompact situation) mainly because the weak* cluster points of the sequence Q^n are always either zero or probability measures in compact or discrete semigroups. In other semigroups, for example in $[0, \infty)$ under multiplication and the usual topology, the sequence Q^n (when Q is the normalized Lebesgue measure on $[0, e]$, ln $e = 1$) vaguely converges to $\frac{1}{2}$ (the unit mass at 0). As can be expected, it is possible to say a good deal more about the sequence Q^n in the general locally compact situation if we assume that the weak cluster points of this sequence are all probability measures.

Theorem 4.3 *Let S be a second countable locally compact Hausdorff semigroup and $Q \in P(S)$. Suppose that the sequence Q^n is weakly conditionally compact. Let $S = \overline{\bigcup_{n=1}^{\infty} S_Q^n}$ and $K = \{\lambda \in P(S): \lambda \text{ is a weak cluster point of } Q^n\}$.*

Let us define

$$S_0 = \overline{\bigcup\{S_\lambda : \lambda \in K\}} \quad \text{and} \quad S_1 = \bigcup\{S_\lambda : \lambda \in K\}.$$

Then S_0 is the closed completely simple minimal ideal of S with a compact group factor. The semigroup S_1 is also completely simple. Here the set K is a group under convolution. Let η be the identity of K. Then we can express S_0 and S_η as

$$S_0 = X \times H \times Y, \quad S_\eta = X \times H_1 \times Y$$

(the usual product representation for completely simple semigroups), where H_1 is a normal subgroup of H. Furthermore, if $\lambda \in K$ and $\lambda \neq \eta$, then

$$S_\lambda = X \times gH_1 \times Y, \quad g \in H - H_1.$$

Clearly, $K = \{\eta\}$ if and only if $Q\eta = \eta$. Therefore, the sequence Q^n converges weakly if and only if there does not exist a normal subgroup H of G such that

$$S_Q(X \times H_1 \times Y) = X \times gH_1 \times Y$$

for some g in $H - H_1$.

Proof Since the sequence Q^n is conditionally weakly compact, by the well-known Prokhorov theorem the sequence Q^n is uniformly tight, so the sequence $(1/n)\sum_{k=1}^n Q^k$ is also uniformly tight. By some standard arguments,

$$\frac{1}{n}\sum_{k=1}^n Q^k \to Q_0 \in P(S),$$

where

$$QQ_0 = Q_0Q = Q_0 = Q_0^2.$$

Let us write $I = S_{Q_0}$. Then I is a completely simple ideal of S with a compact group factor (since Q_0 is idempotent) (see [34]). Let A be any open set containing I. Then we claim that

$$\lim_{n\to\infty} Q^n(A) = 1.$$

Given $\varepsilon > 0$, we choose compact sets K_1 and K_2 with $K_2 \subset I$ such that

$$Q_0(K_2) > 1 - \varepsilon$$

and

$$Q^n(K_1) > 1 - \varepsilon \quad \text{for all} \quad n \geq 1.$$

Since $K_1 K_2 \subset I \subset A$, there exists an open set B containing K_2 such that

$$K_1 B \subset A.$$

Since $Q_0(B) > 1 - \varepsilon$, and $(1/n) \sum_{k=1}^{n} Q^k \to Q_0$, it is clear that there exists k_0 such that

$$Q^{k_0}(B) > 1 - \varepsilon.$$

This means that for all positive integers n,

$$Q^{n+k_0}(A) \geq Q^n(K_1) Q^{k_0}(B) > (1 - \varepsilon)^2.$$

Thus our claim is proven. Now we show that $S_0 = I$. To show this, we write

$$I = \bigcap_{k=1}^{\infty} A_k, \quad A_k \text{ open}, \quad A_k \supset \bar{A}_{k+1}.$$

Let $\lambda \in K$. Let $x \in S_\lambda$. Suppose $x \notin \bar{A}_k$ for some k. Then there is an open set $V(x)$ containing x such that $V(x) \cap A_k = \emptyset$. Since there exists a $\delta > 0$ such that

$$Q^n(V(x)) > \delta$$

for infinitely many n, we must have

$$\lim_{n \to \infty} Q^n(A_k) < 1 - \delta.$$

But this is a contradiction, and therefore, for each k, $S_\lambda \subset \bar{A}_k$ or $S_\lambda \subset I$. Since I is closed, it is clear that $S_0 \subset I$. Since S_0 is an ideal of S, and I is simple, $S_0 = I$.

Now we notice that the set $\overline{\{Q^n : n \geq 1\}}$ (closure = weak* closure) is compact in the weak* topology. By our assumption, it is a subset of $P(S)$. The kernel of this compact abelian semigroup (under convolution) is the set K. Therefore, K is a group (see [34, p. 42]). Let $\eta = \eta^2$ be the identity of K. Then $S_\eta (\subset S_0)$ as well as S_0 are both completely simple semigroups. Let $e = e^2 \in S_\eta$ and let us define

$$X \equiv E(S_0 e), \quad X_1 \equiv E(S_\eta e), \quad Y \equiv E(e S_0),$$
$$Y_1 \equiv E(e S_\eta), \quad H \equiv e S_0 e, \quad H_1 \equiv e S_\eta e.$$

[Here, of course, $E(M) \equiv$ the set of idempotents in M.] Then we can write

$$S_0 \equiv X \times H \times Y, \quad S_\eta \equiv X_1 \times H_1 \times Y_1.$$

If $(x, h, y) \in S_0$, then there exist $(x_n, h_n, y_n) \in S_{\lambda_n}$, where $\lambda_n \in K$, such that $(x_n, h_n, y_n) \to (x, h, y)$ as $n \to \infty$. By the group property of K, there exist $v_n \in K$ such that $\lambda_n v_n = \eta$. This means that $x_n \in X_1$ for all n. Since X_1 is a closed subset of X, $x \in X_1$. Similarly, $y \in Y_1$. Thus we have

$$S_0 \equiv X \times H \times Y \quad \text{and} \quad S_\eta \equiv X \times H_1 \times Y.$$

Now let $\lambda \in K$. Since $\eta\lambda\eta = \lambda$, it is clear that

$$S_\lambda \equiv X \times U_\lambda \times Y,$$

where U_λ is a union of left (or right) cosets of H_1. Since there is a $v \in K$ such that $\lambda v = v\lambda = \eta$, U_λ has to be precisely one left coset of H_1. Thus,

$$S_\lambda \equiv X \times gH_1 \times Y;$$

here, $gH_1 = H_1 g$. Since $\bigcup \{S_\lambda : \lambda \in K\}$ is dense in S_0 and H_1 is compact, it follows that H_1 is a normal subgroup of H.

The proof will be complete if we prove that the set S_1 is also a completely simple semigroup. Notice that if J is an ideal of S_1, then for $x \in J$, $xS_1 \cap S_\lambda \neq \emptyset$ for some $\lambda \in K$; therefore, since K is a group, $xS_1 \cap S_\eta$ is nonempty. This means that

$$S_\eta x S_1 \cap S_\eta \neq \emptyset$$

and therefore, since S_η is simple,

$$S_\eta \subset S_\eta x S_1 \subset J.$$

Since by the structures of λ and η in K, $S_\lambda S_\eta = S_\lambda$ (note that in general, $\overline{S_\lambda S_\eta} = S_\lambda$), we have $S_1 \subset J$. Thus, S_1 is simple. The idempotents in $S_\eta \subset S_1$ are also idempotents in S_0 (which is completely simple) and therefore, primitive. It follows that S_1 is completely simple. Q.E.D.

Now we will discuss the nature of the tail limits of a composition convergent sequence of probability measures on a semigroup. In other words, if $Q_n \in P(S)$ and for all k, w*-$\lim_{n \to \infty} Q_{k,n} = Q'_k \in P(S)$, then a weak* cluster point Q' of the sequence $\{Q'_k\}$ is called a tail limit for the sequence Q_n. In the case when S is a group, there is only one tail limit, which is the normed Haar measure on a compact subgroup. In general semigroups, there can be many tail limits for a sequence in $P(S)$ and the tail limits need not be probability measures. For example, let S be a left zero semigroup and e_n be a sequence in S such that $\lim_{n \to \infty} e_{2n} = e \in S$ and the sequence e_{2n+1} has no cluster point; then, if Q_n is the unit mass at e_n, Q'_k becomes the unit mass at e_k and, clearly, the subsequence Q'_{2k} w*-converges to the unit mass at e while the subsequence Q'_{2k+1} w*-converges to zero.

In the discussion to follow, (Q_n) is a composition convergent sequence of probability measures on a second countable locally compact Hausdorff semigroup S such that

$$\text{w*-}\lim_{n \to \infty} Q_{k,n} = Q'_k \in P(S).$$

Then if Q' is a tail limit in $P(S)$, there exists a sequence (n_i) of positive integers such that

$$\text{w*-lim } Q_{n_i, n_{i+1}} = Q'.$$

Since we have

$$Q_{k, n_i} Q_{n_i, n_{i+1}} = Q_{k, n_{i+1}},$$

it follows by the joint continuity of convolution in $P(S)$ that for each $k \geq 0$,

$$Q'_k Q' = Q'_k.$$

In other words, if Q', Q'' are any two tail limits in $P(S)$ for the sequence (Q_n), then

$$Q'Q'' = Q', \quad Q'Q' = Q'.$$

By a result of Mukherjea and Tserpes (see [34]), the support of each tail limit in $P(S)$ is a completely simple subsemigroup of S. Now let us define the subsemigroup S_0 as the closure of the union of the supports of all those tail limits which are in $P(S)$. Then we claim that S_0 is a completely simple subsemigroup of S. This follows from the following observations:

(i) If Q' is a tail limit in $P(S)$, then for $x \in S_{Q'}$, $xS_{Q'}$ is a minimal right ideal of S_0. To see this, let $y \in S_0$; then there is a sequence $y_n \in S_{Q'_n}$ where each Q'_n is a tail limit in $P(S)$ and $y_n \to y$. Now we have

$$\overline{S_{Q'} S_{Q'_n}} = S_{Q'}.$$

Therefore, if $z \in S_{Q'}$, then

$$zy_n \in S_{Q'} \quad \text{and so} \quad zy \in S_{Q'}.$$

Hence, $xS_{Q'}$ is a right ideal of S_0. Since $S_{Q'}$ is completely simple, $xS_{Q'}$ is a minimal right ideal of $S_{Q'}$ and therefore a minimal right ideal of S_0.

(ii) The union of all such minimal right ideals as in (i) above is a minimal ideal K of S_0; clearly, K is a dense subsemigroup of S_0 and

$$K = \bigcup \{S_{Q'} : Q' \text{ is a tail limit in } P(S)\}.$$

Let e and f be two idempotents in K such that $ef = fe = e$. Then if $e \in S_{Q'} \subset K$ and $f \in S_{Q''} \subset K$, then

$$e = fe \in S_{Q''} S_{Q'} \subset S_{Q''};$$

since $S_{Q''}$ is completely simple, the idempotents in $S_{Q''}$ are primitive. This proves that $e = f$. Thus, every idempotent in K is primitive. In a locally compact topological semigroup, the kernel is closed. It follows that S_0 is completely simple. We now look into the structures of the supports of the tail limits in $P(S)$.

Let Q' be a tail limit in $P(S)$. Let e be an idempotent in its support. Then we write

$$eS_0 e = G, \qquad E(S_0 e) = X, \quad \text{and} \quad E(eS_0) = Y;$$

also

$$eS_{Q'} e = G_0, \qquad E(S_{Q'} e) = X_0, \quad \text{and} \quad E(eS_{Q'}) = Y.$$

It is then known (see Section I) that

$$S_0 = X \times G \times Y \quad \text{and} \quad S_{Q'} = X_0 \times G_0 \times Y_0.$$

(Here "$=$" means "topologically isomorphic.") Let Q'' be any other tail limit in $P(S)$; then

$$\overline{S_{Q'} S_{Q''}} = S_{Q'}, \qquad \overline{S_{Q''} S_{Q'}} = S_{Q''}.$$

It is clear that if $(x, g, y) \in S_{Q'}$, then $(x', g' y) \in S_{Q''}$ for some $x' \in X$ and $g' \in G$. Now $(x, g, y)(x', g', y) = (x, gyx'g', y) \in S_{Q'}$; therefore, $g' \in (yx')^{-1} G_0$. Again, if $y' \in Y_0$, then

$$(x', g', y)(x, g, y') = (x', g'yxg, y') \in S_{Q''};$$

this means that

$$g'(yx)g \in (y'x')^{-1} G_0 \quad \text{or} \quad g' \in (y'x')^{-1} G_0.$$

Hence, $(yx')^{-1} G_0 = (y'x')^{-1} G_0$ whenever $y' \in Y_0$ and $(x', g', y) \in S_{Q''}$. It follows that $S_{Q''}$ is the union of subsemigroups of the form

$$\{(x, g, y) : y \in Y_0 \text{ and } g \in (yx)^{-1} G_0\} = \{x\} \times (y_0 x)^{-1} G_0 \times Y_0,$$

where y_0 is a fixed element in Y_0. We can now sum up the preceding discussion in the following theorem.

Theorem 4.4 *Let S be a locally compact Hausdorff second countable semigroup and (Q_n) be a sequence in $P(S)$ such that for each nonnegative integer k, the sequence $Q_{k,n}$ converges weakly to Q'_k. Let F be the set of all tail limits for the sequence Q_n, i.e., the set of all probability measures on S which are weak cluster points of the sequence Q'_k. Suppose F is nonempty. Then S_0, the closure of the union of the supports of elements in F, is a completely simple subsemigroup of S. Let Q' be a fixed element in F and e be an idempotent in $S_{Q'}$. Let $X_0 \times G_0 \times Y_0$ and $X \times G \times Y$ be the usual product representations of $S_{Q'}$ and S_0, respectively, where $G_0 = eS_{Q'} e$ and $G = eS_0 e$. If Q'' is another element in F, then its support is a union of subsemigroups of the form $\{x\} \times (y_0 x)^{-1} G_0 \times Y_0$, where y_0 is a fixed element in Y_0 and for all y in*

$Y_0, (yx)^{-1} \in (y_0x)^{-1}G_0$. In case $S_{Q'} \cap S_{Q''}$ is nonempty, then there are product representations of $S_{Q'}$ and $S_{Q''}$ of the form

$$S_{Q'} = X_0 \times G_0 \times Y_0, \qquad S_{Q''} = X_1 \times G_0 \times Y_0,$$

where

$$Y_0 X_0 \cup Y_0 X_1 \subset G_0.$$

In case the semigroup S is bicancellative, all the tail limits in P(S) are identical and equal to the normed Haar measure on a compact subgroup. If S is only right cancellative, then the support of each tail limit in P(S) is a left group; and, each tail limit in P(S) is a right invariant probability measure on its support and their supports are all of the form

$$S_{Q'} = X_{Q'} \times G_0,$$

where $X_{Q'} \subset X$, G_0 is a fixed compact subgroup of G, and $X \times G$ is a product representation of the left group S_0.

Now we will study (only briefly) the almost sure convergence (and also, stochastic convergence) of products of independent random variables with values in a semigroup. Earlier, such concepts were studied with respect to their relationship with convergence in distribution for random variables taking values in groups, Banach spaces, discrete completely simple semigroups, and semigroups satisfying certain compactness conditions (well known as conditions (L) and (R)) by a number of authors, including Loynes, Csiszár, Galmarino, Ito–Nishio, Maksimov, Mukherjea–Sun, Byczkowski–Wós and others. (For detailed references, see [34] and [18].) Here we present briefly some results (not complete by any means) which will help us understand these concepts in semigroups, including completely simple nondiscrete semigroups. A more detailed study will be presented elsewhere.

To study the abovementioned concepts, we will assume that S is a locally compact Hausdorff second countable (and therefore, metrizable with a metric d) semigroup, and Q_n a sequence in $P(S)$. Let X_n be a sequence of independent random variables with values in S and distribution Q_n on some probability measure space. Note that then $Z_n = X_1 X_2 \cdots X_n$ is a random variable (measurability of the product follows from the second countable property of S) with distribution $Q_{0,n} = Q_1 Q_2 \cdots Q_n$.

Definition 4.1 The sequence Z_n converges in probability to a random variable Z if

$$\lim_{n \to \infty} P[d(Z_n, Z) < \delta] = 1$$

for every positive δ. (Recall that d is the metric for the topology of S.)

It can be easily verified that when the distributions of Z_n are uniformly tight, then the Cauchy criterion

$$\lim_{m,n \to \infty} P[d(Z_n, Z_m) > \delta] = 0$$

for every positive δ is equivalent to the convergence in probability of the sequence Z_n. It is clear that stochastic convergence of Z_n to Z implies the weak convergence of $Q_{0,n}$ to the distribution of Z. Also, almost sure pointwise convergence of Z_n to some Z implies the stochastic convergence of Z_n to Z. It was first proved by Csiszár [10] that when S is a group, without any nontrivial compact subgroups, the almost sure convergence of Z_n is implied by the weak convergence of the sequence $Q_{0,n}$. The group arguments do not give us any clue regarding the nature of relationship between the above convergence concepts in the case of semigroups. However, it is possible to obtain some interesting results with some probabilistic analysis. An argument similar to that of Mukherjea and Sun (see [34, p. 86]) says that the almost sure limit of Z_n in a completely simple semigroup, when it exists, can be supported only by a left group.

Theorem 4.5 *Let $S \equiv X \times G \times Y$ be a completely simple semigroup. Suppose that Z_n converges almost surely to Z. Then there exists y in Y such that*

$$P(Z \in X \times G \times \{y\}) = 1.$$

Proof Let Q be the distribution of Z. Suppose (x, g, y) is in the support of Q. Let $N(y)$ be any compact neighborhood of y in Y such that

$$Q(X \times G \times N(y)) = \delta > 0.$$

We claim that $\delta = 1$. If not, let ε be any given positive number. Then there exists an open set $V(y)$ containing $N(y)$ such that

$$\delta \leqslant Q(X \times G \times V(y)) \leqslant Q(X \times G \times \overline{V(y)}) < \delta + \varepsilon.$$

We notice that

$$\{Z \in X \times G \times V(y)\} \subset \bigcup_{k=1}^{\infty} \bigcap_{n=k}^{\infty} \{Z_n \in X \times G \times V(y)\}$$

$$\subset \{Z \in X \times G \times \overline{V(y)}\};$$

also

$$Z_n \in X \times G \times V(y) \Leftrightarrow X_n \in X \times G \times V(y).$$

Let us choose integers p and m so large that

$$P\left\{\bigcap_{n=p}^{m} Z_n \in X \times G \times V(y)\right\} < \delta + \varepsilon.$$

Now we can write using independence

$$\delta \leq P\left\{\bigcap_{n=p}^{\infty} Z_n \in X \times G \times V(y)\right\}$$

$$= P\left\{\bigcap_{n=p}^{m} Z_n \in X \times G \times V(y)\right\} \cap \left\{\bigcap_{n=m+1}^{\infty} X_n \in X \times G \times V(y)\right\}$$

$$= P\left\{\bigcap_{n=p}^{m} Z_n \in X \times G \times V(y)\right\} P\left\{\bigcap_{n=m+1}^{\infty} X_n \in X \times G \times V(y)\right\}$$

$$= P\left\{\bigcap_{n=p}^{m} Z_n \in X \times G \times V(y)\right\} P\left\{\bigcap_{n=m+1}^{\infty} Z_n \in X \times G \times V(y)\right\}$$

$$< (\delta + \varepsilon)(\delta + \varepsilon) < \delta^2 + 3\varepsilon.$$

Since $\varepsilon > 0$ is arbitrary, it follows that $\delta \leq \delta^2$ or $\delta(\delta - 1) \geq 0$. This is a contradiction. The theorem now follows. Q.E.D.

Our next result tells us what happens when S is not completely simple, but the double sequence $Q_{n,m}$ ($n < m$, $n \to \infty$) has a nonzero weak* cluster point. We use [5], wherein a similar result was first proved under different conditions.

Theorem 4.6 *Suppose that Z_n converges to Z in probability. Let Q' be a nonzero weak* cluster point of the double sequence $Q_{n,m}$. Then for $y \in S_{Q'}$ and $x \in S_Q$, where Q is the distribution of Z, we have $xy = x$.*

Proof There exists a subsequence (n_i) of positive integers such that $Q_{n_i, n_{i+1}} \to Q'$ as $i \to \infty$. Write $Y_i = X_{n_{2i-1}+1} \cdots X_{n_{2i}}$.

Let $y \in S_{Q'}$ and $y = \bigcap_{j=1}^{\infty} V_j$, where the V_j's are open sets. It is clear that for each j, there are infinitely many i for which

$$Q_{n_{2i-1}, n_{2i}}(V_j) > \delta_j > 0.$$

By an application of Borel–Cantelli lemma, it follows that

$$P\left(\bigcap_{j=1}^{\infty} \bigcap_{k=1}^{\infty} \bigcup_{i=k}^{\infty} Y_i \in V_j\right) = 1.$$

If the event inside the bracket is called A, then for each $w \in A$, there is a subsequence (p_i) of positive integers (depending on w) such that

$$Y_{p_i}(w) \to y.$$

Since we have $Z_{n_{2p_i-1}} Y_{p_i} = Z_{n_{2p_i}}$, it follows that almost surely

$$Z(w)y = Z(w).$$

(Recall that convergence in probability implies almost sure convergence for a subsequence.) It is clear from this that $xy = x$ for $x \in S_Q$, $y \in S_{Q'}$. Q.E.D.

Theorem 4.7 *Suppose that $S \equiv X \times G \times Y$ is completely simple and the sequence Z_n converges to Z in probability. Suppose that there exists a nonzero weak* cluster point of the double sequence $Q_{n,m}$. Then there exists $y \in Y$ such that each such cluster point is supported by the idempotents of $X \times G \times \{y\}$ and $Z \in X \times G \times \{y\}$ almost surely. Also, in this case the set S_0 (see Theorem 4.4), if nonempty, is a set of idempotents in $X \times G \times \{y\}$.*

The proof of the theorem follows at once from Theorem 4.6 and is omitted.

Theorem 4.8 *Suppose that $S \equiv X \times G \times Y$ is completely simple and X is compact. Let H be the closure of the union of the supports of all weak* cluster points of the double sequence $Q_{n,m}$ ($n < m$ and $n \to \infty$). Then Z_n converges to Z in probability if and only if H is a set of idempotents in $X \times G \times \{y\}$ for some $y \in Y$ and $Q_{0,n}$ converges weakly to Q in $P(S)$.*

Proof Note that when X is compact, the set $A^{-1}B$ is compact whenever A and B are compact. As a result, if for all $n > n_0$, $Q_{0,n}(A) > 1 - \varepsilon$ for some compact set A, then since

$$Q_{0,m}(A) = \int Q_{n,m}(x^{-1}A) Q_{0,n}(dx),$$

it follows that for all $m > n > n_0$, we have

$$Q_{n,m}(A^{-1}A) > 1 - 2\varepsilon.$$

This means that each weak* double cluster point of $Q_{n,m}$ is in $P(S)$ and H is nonempty. Also, it is clear that if $H \subset X \times G \times \{y\}$ and $Q_{0,n} \to Q$, then $S_Q \subset X \times G \times \{y\}$.

Note that the "only if" part of the theorem follows from Theorem 4.7. For the "if" part, we assume that H is a set of idempotents in $X \times G \times \{y\}$. Since X is compact and H is a closed subset of the set

$$\{(x, (yx)^{-1}, y) : x \in X\},$$

H is compact. Let $\delta > 0$. Since $S_Q \subset X \times G \times \{y\}$, we can choose a compact set $K \subset X \times G \times \{y\}$ such that $Q(K) > 1 - \delta$. Now each idempotent in $X \times G \times \{y\}$ is a right identity in $X \times G \times \{y\}$. Therefore,

$$zH = z \quad \text{for each } z \text{ in } K.$$

Let us denote by $N_x(r)$ an open neighborhood of x with radius r. For each z in K, let $U_z(H)$ be an open set containing H such that

$$N_z(\delta_z) U_z(H) \subset N_z(\delta), \qquad \delta_z < \delta.$$

There exist $z_1, z_2, \ldots, z_p$ in K such that

$$K \subset \bigcup_{i=1}^{p} N_{z_i}(\delta_{z_i}) \equiv U(K),$$

say. Write

$$U(H) = \bigcap_{i=1}^{+} U_{z_i}(H).$$

Then for $1 \leq i \leq p$,

$$N_{z_i}(\delta_{z_i})U(H) \subset N_{z_i}(\delta).$$

Choose a positive integer k such that

$$P[Z_k \in U(K)] > 1 - \delta$$

and also, for every $m > n \geq k$,

$$Q_{n,m}(U(H)) > 1 - \delta.$$

[Note that $\lim_{n \to \infty} \inf_{m > n} Q_{n,m}(U(H)) = 1$; this can be verified easily.] Now for $n > k$, we have

$$P(d(Z_n, Z_k) < \delta) \geq P\{Z_k \in U(K) \text{ and } X_{k+1}X_{k+2} \cdots X_n \in U(H)\}$$
$$= P(Z_k \in U(K))P(X_{k+1} \cdots X_n \in U(H))$$
$$> (1 - \delta)^2 > 1 - 2\delta.$$

Thus, the Cauchy criterion for convergence in probability is satisfied. The theorem now follows. Q.E.D.

Note that under the conditions of the above theorem, we can also assert that Z_n converges to Z in probability if and only if the set S_0 (see Theorem 4.4) is a set of idempotents in $X \times G \times \{y\}$ for some y in Y and the sequence $Q_{0,n}$ converges weakly. In Theorem 4.8 above, stochastic convergence cannot be replaced by almost sure convergence. This is one essential difference between groups and semigroups in this context. An example demonstrating this behavior is the following:

Take $S = G \times Y$, where $G = \{u\}$ and $Y = \{y_1, y_2\}$ and multiplication is defined as

$$(u, y_1)(u, y_2) = (u, y_2) = (u, y_2)(u, y_2),$$
$$(u, y_2)(u, y_1) = (u, y_1) = (u, y_1)(u, y_1).$$

Then S is completely simple; in fact, it is a right group. Define

$$Q_n(u, y_1) = (n - 1)/n, \qquad Q_n(u, y_2) = 1/n.$$

Then it can be verified easily that

$$Q_{k,n} \to \text{ the unit mass at } (u, y_1)$$

for every nonnegative integer k and the sequence Z_n is convergent in probability, but not convergent with probability one.

We may remark that in some of the above results local compactness can be replaced by a metrizability assumption.

Before we close this section, we discuss very briefly the above convergence concepts for Banach space valued random variables. The well-known classical results on the convergence of sums of independent real random variables due to Lévy, Khinchine, and Kolmogorov extend easily to finite-dimensional random variables. The difficulties in the infinite-dimensional case stem from the fact that bounded subsets of a Banach space may not be relatively compact. K. Ito and M. Nisio [*Osaka J. Math.* 5 (1968), 35–48] overcame these difficulties and proved the following theorem:

Theorem 4.9 *Let (X_n) be a sequence of independent random variables with values in a separable Banach space S. Let μ_n be the distribution of $S_n = \sum_{k=1}^{n} X_k$. Then the following conditions are equivalent:*

(i) S_n *converges almost surely in norm;*
(ii) S_n *converges in probability;*
(iii) μ_n *converges in Prokhorov metric*[7] *(or equivalently, weakly).*

In case the X_k's are symmetrically distributed, then the above conditions are equivalent to the uniform tightness of the measures (μ_n).

Buldygin [4] studied the relationship between almost sure convergence of S_n and uniform tightness of the measures μ_n for random variables, which are not necessarily symmetrically distributed. Let E_0 be a countable dense subset of S. For every $x \in E_0$, there exists y_x^*, a bounded linear functional on S, such that $y_x^*(x) = \|x\|$ and $\|y_x^*\| = 1$. Let us denote the set $\{y_x^* : x \in E_0\}$ by $L(E_0)$. Buldygin proved the following result:

Theorem 4.10 *Let the X_n's, μ_n's, and S be as in Theorem 4.9. Suppose that the measures μ_n are uniformly tight, and $E(X_k) = 0$ (expectation defined by Pettis integral) for all $k \geq 1$. Suppose that for all $x^* \in L(E_0)$, the series $\sum_{k=1}^{\infty} E[x^*(X_k)]^2 < \infty$. Then S_n converges almost surely.*

Buldygin applies his result to a sequence of independent random processes whose realizations belong to $C[0, 1]$ with probability one.

Note 1 Added in Proof In Section III, certain ergodic and convergence properties of stochastic (and bistochastic) chains are singled out and studied briefly with a view to setting up an algebraic point of view for these properties while demonstrating a useful interplay between the study of such chains (in these contexts) and limit theorems for probability measures on semigroups (and groups). To justify the study of this interplay, an easy proof of

---

[7] See Billingsley's book ("Convergence of Probability Measures," p. 238. Wiley, New York, 1968) for a discussion of the Prokhorov metric and weak topology.

the Maksimov theorem (Theorem 3.12) on the convergence behavior of a bistochastic chain is given in Note 3 added in proof (below). It will be shown that this proof extends to most nonhomogeneous stochastic chains, and a proof without using measure theory is not expected to be easier (even when available).

Note 2 Added in Proof Note that the concept of equivalence of nonhomogeneous Markov chains along with the stochastic version of Theorem 3.12 is important in deciding on the convergence of the chain in terms of finding whether a certain simpler chain is ergodic or not.

Note 3 Added in Proof—Another Proof of Theorem 3.12 Let μ_n be a probability measure on the group G of permutation matrices corresponding to P_n. Then there are elements a_n in G (see the proof of Theorem 1 in [6]) such that for all positive integers k,

$$\mu_{k,n_i} \to \lambda_k, \qquad \mu_{k,n}a_n \to \lambda_k z, \qquad \lambda_{n_i} \to \omega_H,$$

where H is a finite subgroup containing z. Now there exists n_0 such that for $n \geq n_0$, $a_n = a_m$ for infinitely many m. Suppose that for some subsequence p_i, $a_{p_i} = b$ (for all i). Then writing $\lim_n P_{k,n} = B_k$ and $\lim_k B_k = B$, we have $Bb = Bz$. Since $z \in H$, $z(C_i) = C_i$, $1 \leq i \leq p$. Let $A = \{x \in G : x(C_i) = C_i, 1 \leq i \leq p\}$. Then A is a subgroup containing H. Since $\lambda_{n_i} \to \omega_H$, there exists k_0 and N ($>n_0$) such that for $n \geq N$, $\mu_{k_0,n}a_n(A) > 2/3$. Since for $n \geq n_0$, $a_n \in A$, this means that $\mu_{k_0,n}(A) > 2/3$ whenever $n > N$. By Lemma 3.2, $\sum_{n=1}^{\infty} \mu_n(G - A) < \infty$. The theorem is now clear. Q.E.D.

We remark that Theorem 1 in [6] is valid in any general compact Hausdorff topological semigroup (second countable) and this makes Theorem 3.12 generalizable (with a similar proof as above) to the case of nonhomogeneous stochastic chains. Finally we state an interesting property of any sequence μ_n of probability measures on a finite group G: Given any subsequence (n_i) of positive integers, there is a subsequence $(n_{i_j}) \subset (n_i)$ such that for all positive integers k, $\mu_{k,n_{i_j}} \to \lambda_k$ and $\lambda_{n_{i_j}} \to \omega_H$, where H is a finite subgroup (depending on the subsequence n_i); also, the λ_k depend on (n_i) and $\lambda_k \omega_H = \omega_H$, and the different subgroups H that correspond to the different subsequences are all conjugate to each other. This property of conjugacy seems to be new, but not very hard to prove.

REFERENCES

1. Birkhoff, G., Tres observaciones sobre el algebra lineal, *Univ. Nac. Tucumán Rev. Ser. A* **5** (1946), 147–150.
2. Brown, D., On clans of nonnegative matrices, *Proc. Amer. Math. Soc.* **15** (1964), 671–674.
3. Brown, G., and Moran, W., Sums of random variables in groups and the purity law, *Z. Wahrscheinlichkeitstheorie und Verw. Gebiete* **30** (1974), 227–234.

4. Buldygin, V. V., On the convergence of series of independent random variables taking on values in a Banach space, *Theor. Probability Math Statist.* No. 6 (1975), 31–38.
5. Byczkowski, T., and Woś, J., On infinite products of independent random elements on metric semigroups, *Colloq. Math.* **37** (1977), 271–285.
6. Center, B., and Mukherjea, A., More on limit theorems for iterates of probability measures on semigroups and groups, *Z. Wahrscheinlichkeitstheorie und Verw. Gebiete* **46** (1979), 259–275.
7. Chatterjee, S., and Seneta, E., Towards consensus; Some convergence theorems on repeated averaging, *J. Appl. Probability* **14** (1977), 89–97.
8. Clark, W. E., Remarks on the kernel of a matrix semigroup, *Czechoslavak Math. J.* **15** (1965), 305–310.
9. Clifford, A. H., and Preston, G. B., "The Algebraic Theory of Semigroups," Vol. 1, Math Surveys, No. 7. Amer. Math. Soc., Providence, Rhode Island, 1961.
10. Csiszár, I., On infinite products of random elements and infinite convolutions of probability distributions on locally compact groups, *Z. Wahrscheinlichkeitstheorie und Verw. Gebiete* **5** (1966), 279–295.
11. DeGroot, M. H., Reaching a consensus, *J. Amer. Statist. Assoc.* **69** (1974), 118–121.
12. Dunford, N., and Scwartz, J., "Linear Operators," Part I. Wiley (Interscience), New York, 1958.
13. Grenander, U., "Probabilities on Algebraic Structures." Wiley, New York, 1963.
14. Grenander, U., "Lectures in Pattern Theory," Vol. 1. Springer-Verlag, Berlin and New York, 1976.
15. Grintsevichyus, A. K., On the continuity of the distribution of a sum of dependent variables connected with independent walks on lines, *Theor. Probability Appl.* **19** (1974), 163–168.
16. Hartman, P., Infinite convolutions on locally compact abelian groups and additive functions, *Trans. Amer. Math. Soc.* **214** (1975), 215–231.
17. Heyer, H., Probabilistic characterization of certain classes of locally compact groups, *Symp. Math.* **16** (1975), 315–355.
18. Heyer, H., "Probability Measures on Locally Compact Groups." Springer-Verlag, Berlin and New York, 1977.
19. Högnäs, G., Random semigroup acts on a finite set, *J. Austral. Math. Soc.* **23** (1977), 481–498.
20. Hofmann, K. H., and Mostert, M. S., "Elements of Compact Semi-Groups." Merrill, Columbus, Ohio, 1966.
21. Huang, C. C., Isaacson, D., and Vinograde, B., The rate of convergence of certain nonhomogeneous Markov chains, *Z. Wahrscheinlichkeitstheorie und Verw. Gebiete* **35** (1976), 141–146.
22. Isaacson, D., and Madsen, R., "Markov Chains: Theory and Applications." Wiley, New York, 1976.
23. Isbell, J. R., Infinite doubly stochastic matrices, *Canad. Math. Bull.* (1962), 1–4.
24. Johansen, S., A central limit problem for finite semigroups and its applications to the imbedding problem for finite state Markov chains, *Z. Wahrscheinlichkeitstheorie und Verw. Gebiete* **26** (1973), 171–190.
25. Kendall, D. G., On infinite doubly stochastic matrices and Birkhoff's problem 111, *J. London Math. Soc.* **35** (1960), 81–84.
26. Kendall, D. G., Extreme point methods in stochastic analysis, *Z. Wahrscheinlichkeitstheorie und Verw. Gebiete* **1** (1963), 295–300.
27. Maksimov, V. M., Necessary and sufficient conditions for the convergence of the convolution of non-identical distributions on an arbitrary finite group, *Theor. Probability Appl.* **13** (1968), 287–298.
28. Maksimov, V. M., Convergence of nonhomogeneous bistochastic Markov chains, *Theor. Probability Appl.* **15** (1970), 604–618.

29. Maksimov, V. M., A generalized Bernoulli scheme and its limit distributions, *Theor. Probability Appl.* **18** (1973), 521–530.
30. Martin-Löf, P., Probability theory on discrete semigroups, *Z. Wahrscheinlichkeitstheorie und Verw. Gebiete* **4** (1965), 78–102.
31. Mirsky, L., Proofs of two theorems on doubly stochastic matrices, *Proc. Amer. Math. Soc.* **9** (1958), 371–374.
32. Mukherjea, A., Limit theorems for convolution iterates of a probability measure on completely simple or compact semigroups, *Trans. Amer. Math. Soc.* **225** (1977), 355–370.
33. Mukherjea, A., and Nakassis, A., On the limit of the convolution iterates of a probability measure on $n \times n$ stochastic matrices, *J. Math. Anal. Appl.* **60** (1977), 392–397.
34. Mukherjea, A., and Tserpes, N. A., "Measures on Topological Semigroups," Lecture Notes in Mathematics, No. 547. Springer-Verlag, Berlin and New York, 1976.
35. Révész, P., A probabilistic solution of problem 111 of G. Birkhoff, *Acta Math. Acad. Sci. Hungar.* **13** (1962), 187–198.
36. Rosenblatt, M., "Markov processes: Structure and Asymptotic Behavior." Springer-Verlag, Berlin and New York, 1971.
37. Seneta, E., "Non-Negative Matrices." Wiley, New York, 1973. [See also his paper in *Studia Math.* **46** (1973), 241–247.]
38. Sun, T. C., On the limit of convolution iterates of a probability measure on 2×2 matrices, *Bull. Inst. Math., Acad. Sinica* (1975).
39. Tortrat, A., Lois de probabilité sur un espace topologique complètement régulier et produits infinis a termes indépendants dans un groupe topologique, *Ann. Inst. H. Poincaré Sect. B* **1** (1965), 217–237.
40. Wolfowitz, J., Products of indecomposable, aperiodic, stochastic matrices, *Proc. Amer. Math. Soc.* **14** (1963), 733–737.

AMS (MOS) 1980 Subject Classifications: 60B99, 60F99

Index

A

Admissible controls, 7, 54
  space of, 7
Admissible trajectories, set of, 10
Algebraic Hilbert space, 80
Atomic formulas, 118

B

Basis of a bistochastic chain, 178
Bochner–Masani duality theorem, 82
Brownian motion, 138

C

Closure theorem, 8
Coin tossing, 135
Consensus problem, 164
Constant, 108
Control measures, 9
Convergence in probability, 195
Correlation function, 75, 79, 98
Correlation functional, 101
Cost function, 4, 11
Csiszár's theorem, 146
Csiszár–Tortrat theorem, 147

D

Density operator, 78, 95

E

Enlargement, 120
Equivalent Markov chains, 180
Equivalent sequences, 112, 120
Ergodic coefficient, 167
External element, 106
External set, 120

G

Girsanov's theorem, 20
Gleason measure(s), 69
  basic, 71
  complex-valued, 72
  E-valued, 71
  Gaussian, 98
  orthogonally scattered, 75
  operator-valued, 83
  positive, 72
  symmetric, 96, 97
  random, 97
    Lévy–Khinchine representation for, 100
  semispectral, 87
  spectral, 83, 85
  weak convergence of, 90, 92, 93
Gleason's theorem, 69
  generalization of, 71

H

Heisenberg picture, 70
Homomorphism, *-quasi, 86
Hyperfine space, 107

I

Indicator measure, 81
Infinitesimal, 114
Integrable
  S-integrable, 131
  uniformly, 131
Internal element, 106
Internal definition principle, 121
Internal set, 120
Itô integral, 141

K

Kalman's matrix Riccati equation, 40, 41
Kernel, 81
Kushner's existence theorem, 4

L

L^*, nonstandard representation of, 127
Lebesgue measure, 134
Lévy–Prokhorov metric, 91
Liftings, S-integrable, 132

M

Maksimov's theorem, 179
Matrices
　backward products of, 164, 166
　stochastic, 147, 152, 156
　strongly ergodic sequence of, 165
　weakly ergodic sequence of, 165
Maximum principle, 32
Measurable spaces, nonstandard representation of, 123
Measure(s)
　control, 9
　indicator, 81
　Lebesgue, 134
　Wiener, 134, 140
Monad, 114, 123

N

n-ary relation, 108
　complement of, 110
Nonstandard extension, 106, 113, 120

O

Observables, 70
Open look controls, 3

P

Permanence principles, 122
Poisson process, 135
Power set, 117
Probability measure(s), convergence theorem for iterates of, 175
　convolution products of, 175
Prokhorov topology, 6
Purity of the limit, 155

R

Random control(s)
　generalized, 9
　measure-valued, 9
Random transition matrix, 32
Random walk, 148
Real numbers
　elementary structure for, 108
　hyperreal, 111
　　finite, 114
　　infinite positive, 114
　nonstandard, 111
　standard, 108
Representative, *-quasi, 86
　normalized, 86
　weakly continuous, 86
Révész' theorem, 183
Rosenblatt's theorem, 146

S

Sample, 126
Schroedinger picture, 70
Semigroup(s), completely simple, 145
　algebraic, 145
　compact, 145
　kernel of, 145, 156
Semigroup property, 32
Sentence(s), 109, 118
　atomic, 109
　simple, 109
　standard, 120
Separation principle, 16, 41
　extension of, 28
Sherstnev's theorem, 71
　for complex-valued Gleason measures, 72
Simple language, 108
Skolem functions, 109, 111
s-operator, 71
　consistent, 91
Standard object, 106
Standard part, 114
Standard part map, 133
Standard real numbers, 108
State, 70
Stationary field of second order, 101
Stochastic differential equation(s), 11, 12, 14, 17, 23, 24, 30, 31, 34, 35, 36, 40, 43, 48, 54, 58, 62, 63
Stochastic functional differential equation(s), 4, 6, 8, 13, 19, 20

Stochastic integral equation(s), 3, 20
Stochastic matrices, semigroup of, 147, 152, 156
Superstructure, 117, 120

T

Term, 109, 118
Terminal condition, 4
Theorem
  Bochner–Masani duality, 82
  Csiszár's, 146
  Csiszár–Tortrat's 147
  Girsanov's, 20
  Gleason's, 69, 71
  Kushner's, 4
  Maksimov's, 179
  Rosenblatt's, 146
  Sherstnev's, 71, 72
  Wigner's, 70, 84
Transfer principle, 106
  simple, 112

V

Vague convergence, 145
Von Neumann algebra, 70
Von Neumann tensor product, 94

W

Weak *-convergence, 145
Wiener measure, 134, 140
Wigner automorphism, 70
Wigner's theorem, 70, 84